Advance Praise

"Jerome does not only master the Lean Six Sigma/Hoshin Kanri theories, but he also lived them. He implemented them in several companies and delivered results. I have had the opportunity to witness it directly, and it was spectacular. Jerome can quickly engage the most skeptical colleagues and align them behind his plan and deliver results. I highly recommend his book, a small investment that will help you deliver great shareholder value. Everybody can benefit from it, great reading."

Patrick Deconinck
3M Senior Vice President, retired

"Jerome Hamilton is truly an expert in understanding how to apply the concepts of Lean across the enterprise while delivering sustainable results. I would highly recommend this book to any leader who is looking to give their enterprise competitive advantage and truly become a Lean leader."

Jon Hartman
Vice President, North American Operations
World Kitchen LLC

"A must-read for those seeking to have a sustainable impact! Jerome Hamilton has been a student of Lean for many years. His unique insights, his learning from the masters of the Toyota Production System and his practical experiences leading and executing in multiple organizations are passionately demonstrated in this work. The Student has become the Coach and Teacher."

Dr. Tracy Joshua
Vice President, Procurement
The Kellogg Company

Make Your Business a Lean Business

How to Create Enduring Market Leadership

Make Your Business a Lean Business

How to Create Enduring Market Leadership

Paul C. Husby

Jerome Hamilton

CRC Press
Taylor & Francis Group
Boca Raton London New York

CRC Press is an imprint of the
Taylor & Francis Group, an **informa** business

A PRODUCTIVITY PRESS BOOK

CRC Press
Taylor & Francis Group
6000 Broken Sound Parkway NW, Suite 300
Boca Raton, FL 33487-2742

CRC Press is an imprint of Taylor & Francis Group, an Informa business

No claim to original U.S. Government works

Printed on acid-free paper

International Standard Book Number-13: 978-1-4398-2999-8 (Paperback)
International Standard Book Number-13: 978-1-1380-8213-7 (Hardback)

Library of Congress Cataloging-in-Publication Data

Names: Husby, Paul C., author. | Hamilton, Jerome, author.
Title: Make your business a lean business : how to create enduring market leadership / Paul C. Husby and Jerome Hamilton.
Description: Boca Raton, FL : CRC Press, 2018.
Identifiers: LCCN 2017008209| ISBN 9781439829998 (pbk. : alk. paper) | ISBN 9781138082137 (hardback : alk. paper)
Subjects: LCSH: Lean manufacturing. | Production management. | Organizational effectiveness.
Classification: LCC TS155 .H859 2018 | DDC 658.5--dc23
LC record available at https://lccn.loc.gov/2017008209

Visit the Taylor & Francis Web site at
http://www.taylorandfrancis.com

and the CRC Press Web site at
http://www.crcpress.com

Contents

Authors

Paul C. Husby completed a 38-year career with 3M including executive management positions as the managing director of 3M Brazil, division vice president of the Abrasives Division, and corporate staff vice president of Manufacturing and Supply Chain Services. Husby's career included a significant number of operational leadership assignments in manufacturing, engineering, and supply chain prior to his executive leadership assignments. He graduated from the University of Wisconsin, Stout, with a BS in industrial technology. Twelve years of his career were spent in international assignments including Belgium, the United Kingdom, and Brazil; he speaks Portuguese as a second language. Currently a supply chain consultant with a passion for Lean manufacturing, he is also involved with Hope Unlimited of Brazil as the vice president of the board of directors. Hope Unlimited is a nonprofit organization that rescues and transforms the lives of Brazilian street children.

Jerome Hamilton, the CEO of Open Therapeutics™, leads the company's scale and global expansion, as well as manages business development, sales, marketing, and investor relations. He joined Open Therapeutics after serving as Stratasys' senior vice president of global operations, before which he was the vice president of LEAN Six Sigma Operations, Corporate Quality and Acquisition Integration for 3M. There, he played a major role in devising strategy and operational plans for the company's largest business group, the Industrial Business Group. He also led the company's Masking and Surface Protection Business with full P&L responsibility.

Hamilton earned his bachelor of general science from Morehouse College; bachelor of science in industrial engineering from Georgia Tech; master of science in engineering management from the University of Detroit, Mercy; and master of science for international logistics from Georgia Tech. He is a graduate of the Advanced Management Program, Harvard Business School.

Hamilton is a member of the Boards of Directors of Goodwill Easter Seals Minnesota, Real Life 101, and V2SOFT. He was listed in the 2014 *Savoy Magazine* list of Most Influential Blacks in Corporate America, and was recognized as one of the Top 50 diversity professionals in the industry—Global Diversity List, supported by *The Economist* (globaldiversitylist.com).

In the coming months, Open Therapeutics will provide to global researchers a crowdsourcing web platform that will host freely available open pharma and medical technologies framework, created by Jason Barkeloo and his team. Scientists can freely adopt the open technologies and develop manuscripts on the science they conduct around the technologies.

About Open Therapeutics LLC

Open Therapeutics is a 7-year old company that curates and develops open medical, biopharma, and synthetic biology-based biotechnologies. Among the technologies being freely opened and crowdsourced to the global community are essential proteins for developing antibiotic and anticancer therapeutics, immunotherapy and oncolytics, and biomarker inducers. The company is headquartered in Cincinnati, Ohio with laboratory operations in Covington, Kentucky, and Madrid, Spain via its strategic partner, Bacmine SL (http://www.bacmine.com).

Introduction

What Will I Learn?

How my company can maximize the probability of being successful for a hundred years or more.

Lean Enterprise, It Is about Having an Effective Operational Management System

Who Needs This Book?

At least 85% of all companies in existence today! Why? Look at history and see the probability of your company existing for the next 50 years. Of the 500 companies listed on the S&P 500 in 1957, only 15% are listed today.[1] Nearly half of the 25 companies included in Peters & Waterman's 1982 book *In Search of Excellence* do not exist today.[2] Is there really any reason to believe that we will not see the same story repeating itself in the future? What should responsible executives be doing to improve their companies' chances of long-term survival? *All companies have an operational management system, but is it continuously improving strategic competitive advantage, increasing operational execution excellence, and lowering operational cost through daily continuous improvement from an army of problem-solving scientists?* If not, company CEOs and COOs should start implementing the Lean Enterprise Operational Management Systems (LEOMS) today as it will transform their companies, increase their value to customers and investors, and improve employee satisfaction. The LEOMS will not replace great strategy for companies with strong sustainable strategic competitive advantage and differentiating competencies, but it will get the most out of every company's strategy, organization, processes, and assets. Lean is best known for some of its methods and practices, often thought of as only relevant to supply chain and factory floor operations. This is a gross misconception and limits the potential of Lean to improve enterprise-wide operational performance. This book's goal is to help you turn your company into a "Lean machine" that will continuously accelerate the rate of improving its competitive advantage through application of the LEOMS, thus creating a more sustainable Lean enterprise. Its purpose is to change the common understanding of Lean from only an operations improvement system to a true enterprise operational management system. Through this book, business leaders will learn about the LEOMS and how to implement it by following the Seven Wave Implementation Process. Becoming a Lean enterprise ensures effective implementation of business strategic goals and priorities by aligning and optimizing all resources and holding annual and 3-year improvement target owners accountable, and by maintaining a "laser focus" on achieving their few critical targets. There appear to be numerous plausible explanations for the absence of understanding the LEOMS's importance in improving performance and its snail-like widespread successful adoption. The six most recognized failure sources are addressed in this book:

1. Three quarters of TQM, reengineering, strategic planning, and downsizing initiatives have failed, most frequently because of the failure to recognize how the company culture would react to required changes for success of the initiative.[3] Lean must become part of the culture to be sustainably successful through vigorous application of the culture change process as described in Chapter 4.

2. There has not been a guidebook with a set of processes, practices, and tools to follow during implementation of the LEOMS to achieve the strategic and operational benefits of continually improving operational and strategic competitive advantage.

3. Many commonly used Lean models such as the Toyota House are valuable but speak to supply chain operations and not the entire enterprise, its distinctive competencies, and sources of strategic competitive advantage.

4. Typical Lean implementations are started in factory operations and built out to customers and suppliers. This makes great sense and gets results, but it also reinforces the perception of Lean as only relevant to supply chain operations. In addition, TPS's models like the Toyota House represent enormous implicit knowledge and a multitude of processes, practices, and tools that need to be made explicit for practitioners to be able to apply and to understand their interrelationships within the whole system.

5. Applying Hoshin Kanri is growing, but still not widely adopted by companies as a component of their strategic and operational business planning and execution management process. Hoshin Kanri resolves many issues such as misalignment and lack of focus. No doubt these are sources of significant waste, but they are invisible wastes,[4] which are hard to measure and quantify. Most companies are so accustomed to norms of cordial anarchy and plan execution failures that these wastes continuously cause frustration but also are viewed as normal.

6. Many academics and university courses teach Lean more as an improvement methodology, sometimes even positioned as comparable to Six Sigma. Very few get to the heart of Lean as the best enterprise operational management system in existence.

Whatever the causes, little real progress has been made in advancing understanding and sustainable success of LEOMS applications. This book is intended to contribute to advancing LEOMS application success. The purpose of this book is to reposition the discussion of Lean from a set of tools and processes to focus on Lean as Taiichi Ohno positioned it in his book *Toyota Production System* as a "system that will work for any type of business."[5] It is the goal of this book to stimulate a new dialogue, repositioning the Lean discussion to focus on the LEOMS. The book introduces new system models as part of an implementation guide to assist in placing the focus on the LEOMS and includes references to great contributions by a multitude of authors during the last 35 years.

Why Should You Read This Book?

Authors Jerome Hamilton and Paul Husby bring more than 60 years of operational management, global business executive, and international experience in businesses in many market segments, and bring their knowledge and perspective to the discussion of creating and sustaining a Lean enterprise.

Chapter Summaries

Chapter 1: The Lean Enterprise Operational Management System Overview

Chapter 1 provides background and chronology of manufacturing operational system development, highlighting the genius of Taiichi Ohno, who, seeing the fragmented pieces, developed a very logical total enterprise operational management system that includes strategic plan deployment, operational plan management, and continuous improvement elements. The chapter also identifies applications of the LEOMS's principles, tools, and thinking to business organization processes outside of plant operations.

Chapter 2: Why Lean Enterprise?

Chapter 2 introduces Lean enterprise and its operational management system, identifying four of the LEOMS's distinctive differentiating attributes that validate it as the best operational management system, with no close contenders.

Chapter 3: Business Strategy-Driven Lean Enterprise

Chapter 3 demonstrates how the Lean enterprise business model continuously drives operational improvement to increase an enterprise's strategic and operational competitive advantage. This is accomplished through rigorous application of the Lean enterprise business model ensuring that enterprise's strategy connects and drives Lean's powerful operational management system to continuously improve strategic, financial, and operational performance.

Chapter 4: Engaging Leadership and Preparing the Culture Change Plan

Chapter 4 deals with the necessity of ensuring that company culture will enable successful creation of an army of continuous improvement scientists. Company culture related to people management and development is the single biggest reason for failure of companies to achieve their LEOMS implementation goals. A basic tenet of cultures is that they only accept change that fits within the sphere of truths as it is interpreted by existing culture beliefs and norms, unless there is a cause and purpose for changing that provides a compelling argument to justify changing existing beliefs and norms. Culture change requires strong, persistent top leadership that builds an organizational army of believers who embrace the change because it is in their personal and the company's long-term best interest.

Chapter 5: Lean Enterprise Transformation Preparation and Launch

Chapter 5 addresses preparation of company leadership to understand the LEOMS, leadership roles within the system, an overview of the seven implementation waves, completion of wave 1: strategic alignment, and 10 launch preparation steps. The chapter walks through the steps to create and deploy effective operational plans assuring implementation management integrity through application of the LEOMS's Hoshin Kanri operational planning and execution management process. Hoshin Kanri is a logical and rigorous process to define 1- and 3-year targets in a process that also aligns organizational priorities to maintain consensus, clarity, and commitment to achieving defined targets. The chapter concludes with an application of wave 1 at MMC, the company we created that will be used in the remaining chapters to provide application examples.

Chapter 6: Lean Enterprise Operational Management System Factory Operations Implementation

Chapter 6 covers waves 2 through 7, including the deliverables and information required for each wave along with application examples at MMC.

Chapter 7: Applying the LEOMS to Enterprise Supply Chains

Chapter 7 covers the application of the LEOMS principles and tools to the wider supply chain. The chapter introduces a tool set called the supply chain operations reference model, which includes all processes required for an effective supply chain. The model is a great aid to quickly identify what specific process needs to be improved to either eliminate failure or improve supply chain performance. MMC application examples are included to assist readers in understanding and applying the LEOMS to supply chains.

Chapter 8: LEOMS Application to Transactional Processes

Chapter 8 addresses applying Lean principles and practices to transactional business processes utilizing swim lane process mapping tools to documenting and improving transactional business processes along with MMC application examples.

Why This Book Matters

The competitive environment continues to intensify.

- Global market access continues to grow as more and more countries around the world participate in the global economy. This trend is likely to continue as underdeveloped countries are recognizing great value through building "free trade"-based economies and in the process reducing poverty. Newer strong players like China, India, and Brazil are asserting themselves both politically and economically. Their strong and growing influence is being felt in the world.

■ Changing customer needs and wants are being greatly influenced by such trends as the desire to have a smaller environmental footprint. Customers are expecting to receive greater value from their purchases. They want them to last longer, be aesthetically pleasing, and be easy to dispose of at the end of their life through recycling or reuse.

■ Developing country product and service advantages will continue as there are still many countries with significant populations that are just entering the global competitive marketplace. Sustaining competitive advantages in developed countries will be very dependent on innovation, optimized use of resources, along with education and training to develop new generations who are prepared with the knowledge and skills required to be employed in this new economy and to rebuild and sustain the middle class, the foundation of a long-term sustainable economy.

■ Increasing and volatile transportation costs will continue to be a great challenge and will complicate decisions such as how to configure your supply chain and where to source products. While it is hard to predict future petroleum cost, it certainly is very likely to continue to increase until countries like the United States become self-sufficient and export enough to influence a free market price for oil and gas along with developing wind and solar energy sources.

■ The trend of shorter product life cycles is likely to continue, as global markets become even more competitive, resulting in greater speed to market of new product ideas and technological innovation.

■ New materials technologies represent a potential for great advantage and a competitive threat. They represent the ability to create products with new functionality and features, while also enabling possible entry into new markets with unique and higher-performing products.

■ Developing countries are producing greatly increasing numbers of engineers and scientists. China alone produces six to seven times as many engineers as the United States. This is likely to be one of the most important of all these trends, as developing countries gain financial and technological know-how and can leverage the tremendous number of engineers and scientists into better, cheaper, and more advanced products.

■ Information technology is transforming business at a speed and scope that has never been experienced in history, creating disruptive changes such as Amazon's real threat to make obsolete retail stores, as they continue to enhance their service capabilities and supply sources along with the growing application of "Big Data," allowing companies to segment and target customers in ways that were not practical in the past.

- Free and open markets in the last three decades have greatly reduced global poverty; in fact, it has been the greatest antipoverty program in history and strengthened relationships among participants. It has also contributed to reducing the purchasing power of middle-class populations in most developed countries, particularly those without high-value job skills and/or university degrees.

The Goals of This Book

- To help readers understand Lean as an enterprise operational management system.
- To supply readers with an implementation guide and knowledge of processes, practices, and tools to implement the LEOMS.
- To assist those who take the challenge of becoming a Lean enterprise to overcome the most common causes of failure or disappointing result by adapting their company culture to be compatible with Lean's people management culture.
- To see the LEOMS as the best enterprise operational business model available in the world and not simply as a set of tools and techniques applied to reduce cost and increase speed.
- To position adoption of Lean as an enterprise transformation versus simply changes to operational practice. This will make an order-of-magnitude difference in results achieved from implementation of the LEOMS.
- To provide business leaders with foundational understanding, a roadmap, and tools to transform their business into a Lean enterprise.

Who Can Benefit from This Book?

- Business leaders of all levels—no other Lean business transformation book available is written *for* business leaders *by* business leaders.
- Strategic planners will benefit by gaining understanding of the LEOMS and how to integrate its practices into their strategic and operational planning processes.
- Academics will be better able to teach the LEOMS based on the system model and explanation of its components as applied to all enterprise functions.

How Should Readers Use This Book?

A roadmap is presented throughout the book, allowing business leaders and strategic planners to follow step-by-step development of their LEOMS implementation plan and its deployment. Concepts and principles are presented in each chapter followed by application examples and author experiences, expert quotations, and comments.

Valuable references are included for readers to develop more understanding of specific tools and practices and their interrelationships. These references will be in shaded text boxes, for example:

> "Therefore, all considerations and improvement ideas, when boiled down, must be tied to cost reduction. Saying this in reverse, the criterion for all decision is whether cost reduction can be achieved."
>
> **Taiichi Ohno**
> *Toyota Production System, p. 53, Productivity Press, 1988*

Author experiences and insights are provided for readers to provide more details about specific tools and practices. These references will be in unshaded text boxes:

> In the mid-1990's I was the Managing Director of 3M Brazil and my leadership team and I spent three weeks during one year with Oscar Motomura at Amana-Keya Leadership Development and learned the value of making invisible cost explicit and discussing them openly.
>
> **Paul Husby**

What Can We Learn by Looking at Industry Laggards and Leaders?

Laggards

What can we learn from the bankruptcy of GM and Chrysler? GM had been in decline for decades. Numerous initiatives were started to transform GM; unfortunately, none of them were finished. There were two initiatives that together held promise to truly transform GM:

1. NUMMI Motors, a joint venture with Toyota and GM as partners. Toyota managed the plant and taught GM Lean, training numerous engineers and managers. In the 30 years since the formation of NUMMI, why had GM not completely transformed their businesses and operations before being forced into bankruptcy? Yes, some plants were transformed but they had 30 years.

2. Saturn was established to set up a new kind of car company. Saturn established a very good brand; customers were loyal and most were part of a Saturn cult. Unfortunately, the status quo won, and this grand experiment was killed.

We have a leadership crisis in many U.S. companies. What has been accomplished during the last 35 years as we started to understand Lean? We have gotten to a point where many Lean tools and practices are applied in U.S. companies. Unfortunately, the understanding of Lean as an enterprise operational management system is practiced in only a very small percentage of companies who proclaim that they are "doing Lean." The real power of Lean is yet untapped in the United States.

Leaders

We believe that in the coming decades, only developed country-based companies that have sustainable strategic advantage and embrace Lean as their operational management system will thrive for 100 years. Companies in North America and Europe will be facing powerful competitors from China, India, and Brazil along with others who will have the technological skills as good as or better than U.S. and European companies. We will see the next round of U.S. company deaths as they fail to grasp and internalize global competitive threats to their survival. It will be a repeat of the 1970s and 1980s when a plethora of traditional U.S. companies ended up on the scrap heap of failed companies. Those choosing to become a Lean enterprise and adopting the LEOMS will improve their probability of success. Maybe a better enterprise operational management system will come along, but until then, following a path laid out by Taiichi Ohno is the only proven path to sustainable operational excellence that supports and enables strategic competitive advantage.

Lean car companies such as Toyota, Honda, and Subaru have a maniacal focus on elimination of all waste and creating customer value. They are consistent leaders in reliability and resale value. Additionally, they are recognized as environmental leaders having reliable products and minimal environmental impact from their products and production processes. Because of their maniacal focus, it is also likely that they will be leaders in improving employee health through changing their lifestyles and preventive medicine practices, thus reducing long-term healthcare costs while reinforcing committed employees. Lean philosophy challenges traditional thinking of treating company, customer, company employee, and community interests as conflicting, believing one must lose for the others to gain. Instead, it accepts this apparent paradox of competing interests and uses it as a catalyst for innovating solutions that allow all interests to be respected. A great example of respecting company and employee interest occurred during the recession of 2007 through 2009; Toyota laid off its contract workers at its Texas truck plant, but instead of laying off permanent workers, they were challenged to use their problem-solving skills to find millions of dollars in savings. Toyota improved its long-term competitiveness and reinforced employee commitment. Fortunately, today most automotive companies are adopting Lean, but one only has to examine current 2016 Toyota operating margins compared to Ford, GM, and Chrysler to see that they still have a long way to go in closing their competitive gap with Toyota.

The proven path to achieving and sustaining competitiveness is becoming a Lean enterprise. Lean is most often viewed as just having to do with manufacturing or transactional operations. It has not been taken on board by business leaders as a business philosophy and system, capable of continuously increasing enterprise value for customers and shareholders, a matter of long-term survival, while building high commitment in company employees.

> "The Toyota Production System, however, is not just a production system. I am confident it will reveal its strength as a management system adapted to today's era of global markets and high-level computerized information system."
>
> **Taiichi Ohno**
> *Toyota Production System, p. xv, Productivity Press, 1988*

So what about Toyota's fall, with multiple recalls and the loss of confidence by their customers? What does this tell us about relevance and value of the Toyota Production System in today's world? In a 2007 interview in the HBR July–August 2007 edition,[6] Katsuaki Watanabe, president of Toyota, made three revealing statements:

- ■ "I realize that our systems maybe overstretched… We must make that issue visible."
- ■ "If a time comes at Toyota when I need to put my foot on the brake pedal rather than on the accelerator, I won't hesitate to do so."
- ■ "It is my job to pull the Andon cord."

Toyota leadership demonstrated that they were human, although Watanabe-san demonstrated that he understood the risk; he did not take any action to set in place early detection signals to tell him that it was time to act. Three years later with embarrassing U.S. market quality and safety issues, the seriousness of the problem was recognized. This failure cost Watanabe-san his position and Toyota hundreds of millions of dollars as they worked to recover their highly esteemed reputation. It was serious enough that Akio Toyoda, Toyota president and CEO, testified before the U.S. Congress on February 23, 2010[7] to explain what had happened at Toyota:

- ■ "Toyota has, for the past few years, been expanding its business rapidly. Quite frankly, I fear the pace at which we have grown may have been too quick."

- "I would point out here that Toyota's priority has traditionally been the following: First, Safety; Second, Quality; and Third, Volume. These priorities became confused, and we were not able to stop, think, and make improvements as much as we were able to before, and our basic stance to listen to customers' voices to make better products has weakened somewhat."
- "We pursued growth over the speed at which we were able to develop our people and our organization, and we should sincerely be mindful of that."
- "I regret that this has resulted in safety issues described in the recalls we face today, and I am deeply sorry for any accidents that Toyota drivers have experienced."

Toyota proved that they were human, as their great success of the past few decades led to a false confidence and resulted in making expansion decisions that violated principles that built their success. The third point in Akio Toyoda's explanation gets to the heart of the matter: Toyota growth outpaced their ability to train personnel sufficiently to be able to work within and sustain the Toyota Production System. It has been passed on to new Toyota operations and people through a process of experienced teachers, with great depth of knowledge, personally coaching new people. This process was typically accomplished one-on-one and went on for extended periods of time. Toyota wanted new people to understand not only the visible dimensions of TPS but also all the invisible, for example, the invisible depth of understanding of how one element of TPS was interconnected to many other elements. In the end, this knowledge transfer process, which created their success, also was a limitation that they did not properly respect. Having read Taiichi Ohno's book many times, I get this! Each time I read the book, I see something new that I previously did not catch or understand. Toyota's recent failures were not caused by a failure of their system but a failure of their leadership. I think the open and candid statements of Akio Toyoda reflect a somewhat unique event and the top man at Toyota took ownership for company failure; it was very refreshing! The bottom line is this: sustainable twenty-first century enterprises will require creating customer value through innovation, speed, service excellence, and competitive cost. The companies who have great innovation and true Lean enterprises will be market leaders, as they will demonstrate speed to market and cost competitiveness, enhancing their value and leveraging their innovation in support of their customers.

Summary

This book is offered with total admiration and respect for all the authors and contributors to TPS–LEOMS over the past 35 years who have helped to open the eyes and educate all of us who have passion for Taiichi Ohno's LEOMS. Its purpose is to create a more visible and complete model of the LEOMS, making it more understandable to business leaders and practitioners, thus contributing to increased strategic and operational benefits of becoming a Lean enterprise. Taiichi Ohno, in his book *Toyota Production System*,[5] challenged all of us TPS believers to contribute to strengthening his system. This book is offered in the spirit of his challenge.

Final note: Throughout the rest of the book, the *Lean Enterprise Operational Management System* will be referred to by its complete name and the acronym *LEOMS*.

References

1. Gary Cokins, Why do large companies fail? *Analytics*, June 2012.
2. Tom Peters and Robert Waterman, *In Search of Excellence*, New York: HarperCollins Publishers, 1982.
3. Kim S. Cameron and Robert E. Quinn, *Diagnosing and Changing Organizational Culture*, Hoboken, NJ: John Wiley & Sons, 2011.
4. Oscar Motomura, *Managing the Invisible*, São Paulo, Brasil: Amana Key, 1998.
5. Taiichi Ohno, *Toyota Production System*, New York: Productivity Press, 1988.
6. Katsuaki Watanabe, President of Toyota, interview by the *Harvard Business Review*, July–August 2007.
7. Akio Toyoda, President and CEO, Toyota, U.S. Congressional Testimony, February 23, 2010.

Chapter 1

The Lean Enterprise Operational Management System Overview

What Will I Learn?

Gain a high level understanding of LEOMS and its background.

Origin and Evolution of Lean

Since World War II (WWII), rapid increases in global market access have brought accelerating changes in the global competitive environment. These competitive market forces have also quickened the pace of change in manufacturing management thinking and practices. Every generation for more than 200 years has made contributions to the evolution of manufacturing technology and management systems (see Figure 1.1). These contributions include fifteen major management tools, practices, and methodologies along with eight enabling technology innovations, many of which contributed to the Lean Enterprise Operational Management System (LEOMS) development that is recognized as the best operational management system known to date. It is being adopted by enterprises around the world.

The Story

1. The first major contribution to Lean manufacturing practice was Eli Whitney's innovation (1765–1825). An American inventor, pioneer, mechanical engineer, and manufacturer, Eli Whitney is best remembered for his cotton gin invention. He also affected U.S. industrial development by manufacturing government muskets in 1799. He translated interchangeable parts concepts into a manufacturing system, giving birth to American mass-production systems. Whitney saw an opportunity to become rich and

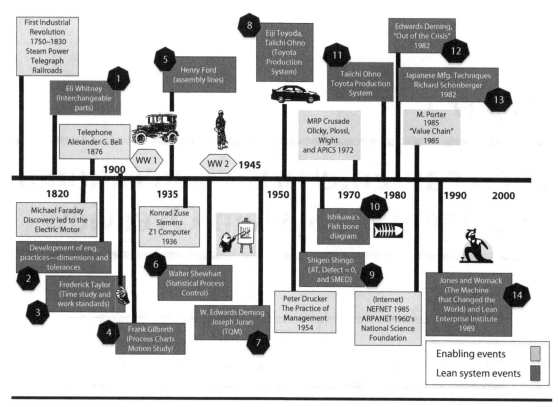

Figure 1.1 Manufacturing management system chronology.

increase Southern prosperity if a machine to clean seeds from cotton could be invented. He set to work at once and within days had drawn a sketch to explain his idea; 10 days later, he constructed a crude model that separated fiber from seed. His gin, however, was a minor accomplishment compared to perfecting interchangeable parts. Whitney developed a musket design with interchangeable parts after taking a U.S. Army contract to manufacture 10,000 muskets at a price of under $14.00 each.

2. Over the next 100 years, manufacturers primarily focused on specific engineering conventions, practices, and process technologies. During this time, our engineering drawing systems, modern machine tools, and large-scale processes, such as the Bessemer process for making steel, were all developed. Few people concerned themselves with movement of products from one discrete process to another or with logistics systems from suppliers, within factories, and to customers. No one was asking questions such as
   a. What happens between processes?
   b. How should multiple processes be arranged within a factory?
   c. How does the chain of processes function as a system?
   d. How does each worker's task relate to other workers?
      This began to change in the late 1890s when Industrial Engineering was developing as an important technology discipline.

3. The first and best-known industrial engineer is Frederick Winslow Taylor, the father of scientific management. He was born into an upper class,

liberal Philadelphia family on March 20, 1865 and died in 1915. Taylor was a compulsive adolescent and was always counting and measuring things to figure out a better way of doing something. At age 12, he invented a harness so he could not sleep on his back, hoping to avoid having nightmares. At age 25, Taylor earned an engineering degree at Stevens Institute of Technology in New Jersey while employed full time. He had a degree from an exclusive university and excelled in math and sports, but chose to work as a machinist and pattern maker in Philadelphia for Enterprise Hydraulic Works. After his hydraulic works plant apprenticeship, he became a shop floor operator at Midvale Steel Company. He started as a shop clerk and quickly progressed through positions of machinist, foreman, maintenance foreman, chief draftsman, director of research, and finally chief engineer. While working there, he introduced a piecework compensation plan at Midvale Steel Company's factory. He closely watched how work was done and measured the quantity produced to find the most efficient way to perform specific tasks. Taylor believed that finding the right challenge for each person and paying him well for increased output was key to improved productivity. At Midvale, he used time studies to set daily production quotas. Incentives would be paid to those reaching their daily goal. Those who did not reach their goal received much lower pay rates. Taylor doubled productivity using time study, systematic controls and tools, functional foremanship, and a new wage scheme, which paid shop floor operators for production output rather than for hours worked. Taylor began to analyze individual workers using work analysis, time study, and standardized work methods. His scientific management approach made a significant contribution by applying science to management, but that success was limited because he never recognized the importance of human factors and behavioral sciences to productivity.

4. Frank Bunker Gilbreth was born in 1868 in Fairfield, Maine and died in 1924. Over the course of his career, he was a bricklayer, building contractor, and engineering manager. He was a member of ASME and the Taylor Society, and a lecturer at Purdue University. His spouse Lillian Evelyn Moller-Gilbreth was born in 1878 in Oakland, California and died in 1972. She graduated with BA and MA degrees from the University of California and went on to earn a PhD from Brown University. She was a member of ASME and also lectured at Purdue University. Frank and Lillian Gilbreth were associates of Frederick Taylor. They had experience in manufacturing industries, which led to an interest in work standardization and methods study. Frank worked at construction sites and noticed that no two bricklayers used identical methods or sets of motions. He improved existing methods by eliminating all wasted motions and raised output from 1000 to 2700 bricks per day. From these studies, Gilbreth developed laws of human motion, which evolved into principles of motion economy. They coined the phrase "motion study" to cover their field of research, thus distinguishing it from those involved

in "time study." The second method developed was process charts, which focuses attention on all work elements, including those non-value-added elements, with the goal of optimizing operator motion and method. Gilbreth's two methodologies were important contributions to industrial engineering's body of knowledge.

5. Henry Ford was born in 1863 in Wayne County, Michigan and died in 1947. He was a pioneering automotive engineer who held many patents on automotive mechanisms. He served as vice president of the Society of Automotive Engineers, which was founded in 1905 to standardize U.S. automotive parts. Ford is best remembered for devising moving factory assembly lines and a completely integrated supply chain. His approach to production revolutionized manufacturing industries by greatly increasing productivity and quality of products. Ford showed an early interest in mechanics, constructing his first steam engine at 15 years of age. In 1893, he built his first internal combustion engine, a small one-cylinder gasoline model, and in 1896, he built an automobile. In 1903, Ford established Ford Motor Company and served as the president from 1906 to 1919 and again from 1943 to 1945. Around 1910, Ford and his right-hand man, Charles E. Sorensen, fashioned the first comprehensive manufacturing strategy. They arranged all manufacturing system elements—people, machines, tooling, and products—into a continuous system for manufacturing Ford's Model T automobile. This strategy was so incredibly successful that he quickly became one of the world's richest men by providing cars everyone could afford. Ford is considered by many to be the first practitioner of just-in-time (JIT) and synchronized manufacturing. Ford's success inspired many others to copy his methods. When the world began to change, Ford's system began to break down, but Henry Ford refused to change the system. Ford's production system depended on laborers working at jobs without meaningful content, causing a loss of dignity and decreased moral. In addition, annual model changes and product proliferation due to multiple colors and options put strains on Ford's system and were not compatible with Ford factories. This became clear when Alfred P. Sloan at General Motors introduced a new strategy. He developed business and manufacturing strategies for managing very large enterprises and dealing with variety. By the mid-1930s, General Motors had surpassed Ford and dominated the automotive market. Many elements of Ford production

ADMIRATION FOR HENRY FORD

"I, for one, am in awe of Ford's greatness. I think that if the American king of cars were still alive, he would be headed in the same direction as Toyota."

Taiichi Ohno
Toyota Production System, Productivity Press, 1988

were sound and still used today. Ford methods were a deciding factor in the Allied victory of WWII.

Henry Ford hated war and refused to build armaments long after war was inevitable; however, when Ford plants finally retooled for war production, they did so on a fantastic scale as epitomized by their Willow Run Bomber plant that built "a bomber an hour." Allied victory supported by massive quantities of material caught the attention of Japanese industrialists. Ford and his system were studied rigorously and much admired by Taiichi Ohno.

6. Rigorous statistical analysis tools have been used in manufacturing by process and quality engineers since Walter Shewhart introduced statistical process control in the 1930s. His work formed the basis of Six Sigma's statistical methodology.

7. Edwards Deming and Joseph Juran introduced thinking and practices, giving birth to Total Quality Management (TQM). Their approaches were embraced by Japanese manufacturing after WWII, and Japan's highest prize for quality is named The Deming Prize. Three decades later, American manufacturers began to accept and apply their thinking and methods.

8. At Toyota Motor Company, Taiichi Ohno and his colleagues studied Ford's production system and American supermarket replenishment methods. These processes inspired Toyota's synchronized manufacturing and JIT techniques. Ohno was founder of Toyota's Production System and creator of the most powerful Enterprise Operational Management System to this day.

Toyota people recognized the contradictions and shortcomings of the Ford system, particularly with respect to employees. General Douglas MacArthur actively promoted labor unions during the U.S. occupation years, making Ford's harsh attitudes and demeaning job structures unworkable in post-war Japan. They were only workable in America because of the "Greatest Generation"—people who had suffered through the Great Depression made Ford's system work in spite of its defects. Toyota soon discovered that factory workers had far more to contribute than just muscle power, as they perform

---

THE TOYOTA PRODUCTION SYSTEM WAS BORN OF NEED

"The Toyota Production System evolved out of need. Certain marketplace restrictions required production of small quantities of many varieties under conditions of low demand, a fate the Japanese automobile industry had faced in the postwar period. The most important objective has been to increase production efficiency by consistently and thoroughly eliminating waste. This concept and the equally important respect for humanity are the foundation of the Toyota Production System."

Taiichi Ohno
Toyota Production System, Productivity Press, 1988

the only value-adding positions in a manufacturing business. This belief led to many practices such as quality circles, team development, and cellular manufacturing.

9. Shigeo Shingo, a Toyota consultant, was asked by Ohno to work on long setup and changeover times of large press operations. Reducing setups to minutes and seconds allowed small batches nearly achieving continuous flow, allowing Toyota to produce a mix of products very efficiently. By 1970, Dr. Shingo perfected his revolutionary concept in manufacturing called "Single Minute Exchange of Dies."[1] This concept was integrated into the Toyota Production System (TPS) and resulted in a significant reduction in operating costs. Shigeo Shingo made numerous contributions to Lean through teaching, consulting, and writing books about it. Dr. Shingo's expertise was a result of his vast experience and knowledge in what can be called modern-day industrial engineering. In 1951, when he first encountered "statistical quality control" concepts, he immersed himself into researching close to 300 companies to gain a better understanding of the subject. By 1959, Dr. Shingo had gained notable fame as an "engineering genius" from his work in developing JIT, contributing significantly to the TPS. In addition, by focusing on production rather than management alone, he was able to establish himself as the world's leading thinker on industrial engineering. One highlight of his approach toward efficient manufacturing can be shown in his work with Mitsubishi Heavy Industries from 1956 to 1958. While working with Mitsubishi, he was able to reduce setup time of ship hull assembly on a 65,000-ton supertanker from 4 to 2 months, setting a new record in shipbuilding. In 1961, Dr. Shingo incorporated his knowledge of quality control to develop the "0 Defects" concept, commonly known as poka-yoke or mistake proofing. This concept was successfully applied to various plants, setting defect-free production records of over 2 years in some operations. All of this took place between about 1949 and 1973. When productivity and quality gains from applying his processes became evident to the outside world, American executives traveled to Japan to study it. They brought back mostly superficial aspects like kanban cards and quality circles. Most early attempts to emulate Toyota failed because they were not integrated into a complete system, as few understood its key underlying principles.

10. Kaoru Ishikawa developed diagrams as he pioneered quality management processes at Kawasaki shipyards and, in the process, became a founding father of modern quality management. Ishikawa diagrams were first used in the 1960s and are considered one of the seven basic tools of quality management, along with histograms, Pareto charts, check sheets, control charts, flowcharts, and scatter diagrams. Known as a fishbone diagram, its shape is similar to the side view of a fish skeleton.

11. Toyota's Production System is the most significant advance since Henry Ford's fully integrated system paced by a moving assembly line.

THE TOYOTA PRODUCTION SYSTEM WAS BORN OF NEED

The Toyota Production System—Relevant in Global Markets and High Tech Information Systems

Taiichi Ohno began developing the Toyota Production System (TPS) in 1949. He and his colleagues worked tirelessly for more than 20 years to fully develop and deploy Toyota Production System's processes and practices across the company and their key suppliers. "The Toyota Production System, however is not just a production system. I am confident it will reveal its strength as a management system in today's era of global markets and high-level computerized information systems."

Taiichi Ohno
Toyota Production System, Productivity Press, 1988

12. In his book *Out of the Crisis,*[2] quality improvement guru W. Edwards Deming recognized that highly successful companies make improvement in the context of their entire supply chain network. He also introduces his 14 points, which elucidate many of The Lean Enterprise Operational Management System's fundamental principles. Deming was an American statistician who is credited with the rise of Japan as a manufacturing nation and the invention of TQM. Deming went to Japan just after WWII to help set up a census of the Japanese population. While he was there, he taught "statistical process control" to Japanese engineers—a set of techniques that allowed them to manufacture high-quality goods without expensive machinery. In 1960, the Japanese Emperor awarded him a medal for his services to Japan's industry. Deming returned to the United States and spent some years in obscurity before the publication of his book *Out of the Crisis* in 1982. In this book, Deming set out 14 points, which he believed, if applied to U.S. manufacturing industry, would save the United States from industrial doom at the hands of the Japanese. Although Deming does not use the term Total Quality Management in his book, he is credited with launching the movement. Most of the central ideas of TQM are contained in *Out of the Crisis*, and his 14 points are key to understanding Deming's thought on eliminating variation. Deming saw variation as a disease threatening U.S. manufacturing. More variation—in part dimensions, delivery times, prices, and work practices—equals more waste, he reasoned. His ideas seemed too radical for most American manufacturing executives at the time, but they are clearly based on Lean's success; Deming was a genius.

13. During the 1980s, North America learned about Toyota's use of JIT manufacturing and kanban from Richard Schonberger's book *Japanese Manufacturing Techniques.*[3] Schonberger traveled to Japan and studied many

TPS practices. He was a true pioneer in getting American companies to begin using these practices.

14. Lean manufacturing gained implementation traction in 1989 with the publication of James Womack and Daniel Jones's book *The Machine That Changed the World*.[4] James P. Womack is the founder and president of the Lean Enterprise Institute, a nonprofit educational and research organization chartered in 1997 to advance a set of ideas known as Lean production and Lean thinking. The book is a 5-year study of the global automobile industry. Their major finding was Toyota's significant competitive advantage over all other competitors based on their TPS. This was the most important contribution to Lean adaptation in America; its well-documented study provided irrefutable proof that the Toyota Production System provided sustainable superior quality and cost results versus all other industry competitors. There are many valuable books and implementation manuals that contributed to our understanding of the TPS's supply chain design and operations, and 40 of them are included in Appendix I.

Lean System

In his book, *Toyota Production System*[5] (translated and made available in English through Productivity Press, 1988), Taiichi Ohno (Toyota engineer and manager, later Executive Vice President of Toyota Motor Company) tells the TPS story. Toyoda Kiichiro, the founder of Toyota Motor Company, was challenged by his father Toyoda Sakichi to start an automobile manufacturing company. When WWII ended, Toyoda Kiichiro discovered that American automobile manufacturers were nine times more productive than Toyota. He, in turn, challenged Taiichi Ohno to catch up with America in 3 years. Japanese market auto sales were much smaller than U.S. market sales, making this challenge even greater. This meant that Ford's inflexible production system would not be practical in producing Japan's high-variety low-volume market requirements. These three leaders recognized that to become competitive would require eliminating all waste. This stimulated their radical thinking, which led to TPS values, principles, and practices to eliminate all waste in pursuit of perfection.

EXAMINE THE DETAILS AND THE BIG PICTURE

"Improving efficiency only makes sense when it is tied to cost reduction. Look at the efficiency of each operator and of each line. Then look at the operators as a group and then at the efficiency of the entire plant (all lines)."

Taiichi Ohno
Toyota Production System, Productivity Press, 1988

Toyota not only one of the world's largest vehicle manufacturing company but also has higher gross margins than all American carmakers. The TPS (LEOMS) is now part of Toyota's DNA, but it has taken decades to perfect it. The more one understands about the LEOMS, the more one marvels at Ohno's TPS. It is difficult to understand how they know "the footsteps of every operator every day," yet maintain employee relations environment positive enough for Toyota plants to remain union-free in the United States. This is one of the many counterintuitive lessons to be learned from a serious study of the LEOMS. How do they continue to improve after 50 years of using the same system? Ohno's book also illuminates the people and experiences that influenced his thinking and the values leading to the LEOMS. The TPS is often represented by the Toyota House (see Figure 1.2) to illustrate some of its core principles, practices, and end purpose. TPS's purpose is achieved by elimination of eight specific forms of waste:

1. Overproduction—making more than is needed
2. Transport—excessive movement of materials
3. Motion—inefficient movement of people
4. Waiting—underutilization of people
5. Inventory—material laying around unused (often a symptom of another waste)
6. Overprocessing—manufacturing to a higher quality standard than expected by the customer
7. Defect correction—time spent fixing defects, including the part that gets thrown away and the time it takes to make the product correctly
8. Underutilized talent and skills

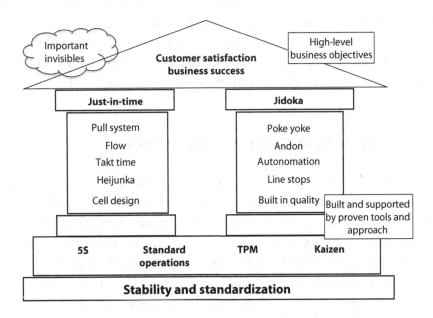

Figure 1.2   Toyota Lean house.

A REVOLUTION IN CONSCIOUSNESS

"A revolution in consciousness is indispensable. There is no waste in business more terrible than overproduction. Why does it occur? Industrial society must develop the courage, or rather the common sense, to procure only what is needed when it is needed and in the amount needed. This requires what I call a revolution in consciousness, a change of attitude and viewpoint by business people. Holding a large inventory causes the waste of overproduction. It also leads to an inventory of defects, which is a serious business loss. We must understand these situations in-depth before we can achieve a revolution in consciousness."

Taiichi Ohno
Toyota Production System, Productivity Press, 1988

Taiichi Ohno correctly believed that improving an integrated value stream was the best approach to deliver value to customers and other supply chain participants. The LEOMS has proven its timelessness by delivering results for more than 50 years and remains the best practice for manufacturing plant and supply chain operations. The Toyota House and eight wastes provide important basic information about the TPS, but they do not provide a complete model to assist in understanding the entire system. In addition, the name Toyota Production System could be interpreted to infer that it only has application to producing products. This was clearly not the intent of the founder, Taiichi Ohno. In his book, he describes it as a business management system well suited to success in any business and global market. This system we have called Lean is intended as a complete enterprise operational management system, starting from the founder's values and principles to its processes and detailed tools and practices. The LEOMS model as represented in Figure 1.3 provides a more complete picture of the LEOMS and will be explored in detail in the remaining chapters of the book.

LEOMS—World's Best Operational Model

The LEOMS is presented in the spirit of Taiichi Ohno's challenge to all future practitioners, "to contribute to the advancement of the Toyota Production System," in this case making the system's depth more visible and understandable for those who wish to adopt it. As not everyone can have a long experienced Sensei to teach and mentor them, the Lean Operational Management System model presented makes the implicit holistic depth of the "Toyota House" explicit. This model was constructed based on significant research of Taiichi Ohno's and Shigeo Shingo's books and complemented by more contemporary Lean System Senseis who have provided more depth of understanding of specific system

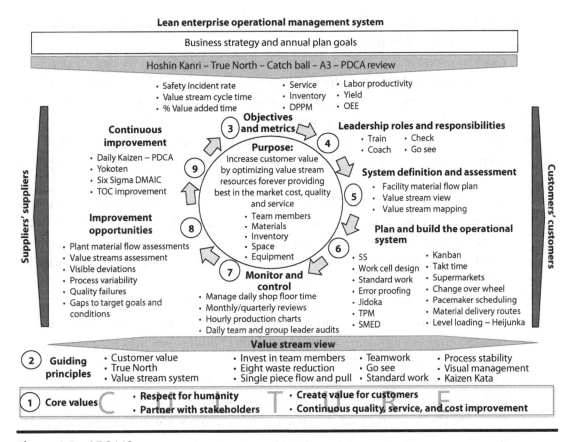

Figure 1.3 LEOMS.

practices and tools. Each time one reviews Taiichi Ohno's *Toyota Production System* and Shigeo Shingo's *A Study of the Toyota Production System,*[6] a new depth of insights into their genius is gained. This system representation defines LEOMS's purpose in the center circle and how it is achieved through the nine components outside the circle.

1. The *LEOMS core values* have been in place for decades at Toyota, and unlike many companies, these values are imbedded in their operational model and are evident in daily activities performed by their leaders and team members. The greatest challenge for companies who seek to adopt the LEOMS is having a culture that will embrace these values, particularly Respect for Humanity as it relates to leading team members, and integrating behaviors into leaders' daily routine until they become "The way we do things around here."

2. *Guiding principles: The LEOMS Guiding Principles* are twelve core system fundamentals starting with a focus on delivering *customer value* and shareowner returns by achieving "True North," LEOMS's vital few short- and long-term targets that must be achieved through an implied contract or bond among all organizational members. The *value stream system* provides a logical system of processes that make same or similar products, extending from suppliers through to end customers in order to create a defined system that

can be focused on and improved. These principles are the LEOMS's rules of the road, the LEOMS way of working.

a. *Customer value* is a criterion applied to all enterprise activities to determine whether they are value-adding or non-value-adding. A Lean enterprise focuses on increasing value received by customers while reducing cost, resulting in improving profits.

b. *True North* keeps an organization focused and driven by a critical few metrics determined from Hoshin Kanri, an annual operational planning process.

c. *Value stream systems* are defined as materials and information flow from supplier's suppliers to end-user customers. Value streams are systems that must be continually improved by reducing waste. Decisions on improvement are made based on optimizing system performance, not individual process performance.

d. *Invest in team members* as a vital source of creating and sustaining defect-free process execution and kaizen, its continuous improvement process that develops an army of problem-solving scientists improving their processes each day.

e. The *eight wastes* provide clarity in identifying system waste and making it easy to find value stream improvement opportunities.

f. *Single piece flow and pull* defines a value stream's ideal flow state including what triggers flow. If only one unit is being worked on at each process with no other inventory and nothing is made until customers consume a unit, the value stream is operating at perfection. This results in value streams with the shortest cycle time and zero waste, the ultimate goal for every Lean value stream.

g. *Teamwork* is essential to Lean success. Teamwork with suppliers, shop floor associates, team leaders, managers, and support resources is an essential organizational attribute. Lean excellence depends on continual closed-loop communications across, up, and down value stream organizations. This will only become a reality if teamwork is practiced as the way of working.

h. *Go see* means spend time observing value stream processes and not making assumptions or creating theories sitting in conference rooms or offices. Reality is on the shop floor observing what happens; it is where waste can be seen and effective ideas for system improvement are born.

i. *Standard work* creates operational consistency from one cycle to another, a foundational element of process stability. It involves defining workstation layout, steps to execute a process, process cycle time, and standard inventory.

j. *Process stability* is the state of a process producing consistent results within a consistent cycle time. It is a basic requirement for building continuous process improvement.

k. *Visual management* is a powerful principle with connections to many Lean system dimensions. Lean system operations are to be transparent so everyone involved can see and understand what is happening.

l. *Kaizen Kata* or continuous improvement daily routines is LEOMS's improvement engine and catalyst. This relentless application of continual system improvement powers Lean systems and stimulates creative problem-solving thinking in pursuit of perfection.

3. *Hoshin Kanri* is the process that establishes *True North*, the vital few metrics guiding a business organization's focus and effort. The shop floor operational metrics are subordinate to these vital few metrics and are built to measure, control, and improve shop floor operations.

4. *Leadership roles and responsibilities* describe key tasks to enable the system to perform such as train, go see, check, and coach team members executing their assigned tasks to control and continuously improve the processes utilizing their full human potential.

5. *System definition, design, and assessment.* Value stream mapping is used to define, design, and assess value streams to continuously improve them.

6. *Plan, build, and operate the value stream system* by applying Lean practices and tools.

7. *Monitor and control* processes to detect deviations from standards, target conditions, and system performance goals.

8. *Improvement opportunities* result from continual value stream assessments, monitoring, and control activities focused on detecting defects, deviations, and gaps to achievement of target conditions and performance.

9. *Continuous improvement* in defining and implementing countermeasures through daily applications of PDCA and Yokoten, and, as appropriate, application of Six Sigma DMAIC and Theory of Constraints methods.

While the LEOMS model presented is a total system representation, it must be acknowledged that recognizing total understanding of Lean in all of its dimensions and contributing to system improvement will be a lifelong pursuit. This is how it should be for those of us who are committed Lean practitioners as the curiosity, pursuit of perfection, and a better way were practiced and preached by Taiichi Ohno.

LEAN PRACTITIONERS MUST CONTINUE TO IMPROVE KANBAN

"It is the duty of those working with Kanban to keep improving it with creativity and resourcefulness without allowing it to become fixed at any stage."

Taiichi Ohno
Toyota Production System, Productivity Press, 1988

Lean System Values

In addition to being foundational beliefs that underpin the LEOMS culture, Lean's values and principles are key to understanding and successfully applying the LEOMS. They reflect the founder's business philosophy, the values to live by in business. There are six decades of validation based on success of companies who have adapted and internalized Lean. They provide a philosophical "True North," boundaries within which the LEOMS's purpose, practices, and tools are successfully applied. Lean has four core values discoverable from Taiichi Ohno's book, *Toyota Production System*, forming the system's philosophical beliefs or tenets (see Figure 1.4).

LEOMS Purpose

Lean enterprises must adapt a waste consciousness and sensitivity, striving daily for zero waste in all organizational activities and at all organizational levels (see Figure 1.5). Waste elimination leads to greater speed through reduced cycle times in all processes both operational and transactional. A significant consequence of waste elimination is radically reduced cost. In addition, this maniacal focus on waste elimination, driven by a pursuit of zero waste, also stimulates break-through innovations in practices and processes. An example is eliminating inventory, which catalyzed practices like level scheduling, mixed model manufacturing,

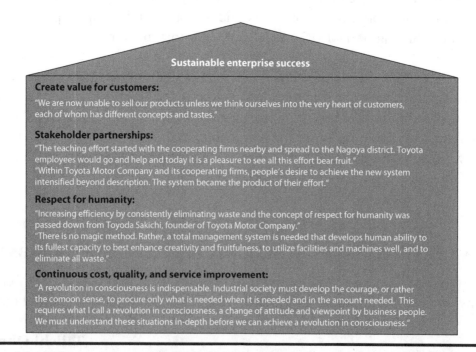

Sustainable enterprise success

Create value for customers:

"We are now unable to sell our products unless we think ourselves into the very heart of customers, each of whom has different concepts and tastes."

Stakeholder partnerships:

"The teaching effort started with the cooperating firms nearby and spread to the Nagoya district. Toyota employees would go and help and today it is a pleasure to see all this effort bear fruit."
"Within Toyota Motor Company and its cooperating firms, people's desire to achieve the new system intensified beyond description. The system became the product of their effort."

Respect for humanity:

"Increasing efficiency by consistently eliminating waste and the concept of respect for humanity was passed down from Toyoda Sakichi, founder of Toyota Motor Company."
"There is no magic method. Rather, a total management system is needed that develops human ability to its fullest capacity to best enhance creativity and fruitfulness, to utilize facilities and machines well, and to eliminate all waste."

Continuous cost, quality, and service improvement:

"A revolution in consciousness is indispensable. Industrial society must develop the courage, or rather the comoon sense, to procure only what is needed when it is needed and in the amount needed. This requires what I call a revolution in consciousness, a change of attitude and viewpoint by business people. We must understand these situations in-depth before we can achieve a revolution in consciousness."

Figure 1.4 TPS core values. (From Taiichi Ohno, *Toyota Production System*, Productivity Press, 1988.)

Key	Zero-waste goal	Relation to profit
1. Customer focus	Zero customer dissatisfaction	Customer input and feedback assures quality. Customer satisfaction supports sales.
2. Leadership	Zero misalignment	Direction and support for development improves cost, quality, and speed.
3. Lean organization	Zero bureaucracy	Team-based operations reduce overhead by eliminating bureaucracy and ensuring information flow and cooperation.
4. Partnering	Zero stakeholder dissatisfaction	Flexible relationships with suppliers, distributors, and society improve quality, cost, and speed.
5. Information architecture	Zero lost information	Knowledge required for operations is accurate and timely, thus improving quality, cost, and speed.
6. Culture of improvement	Zero waste creativity	Employee participation in eliminating operations waste improves cost, quality, and speed.
7. Lean production	Zero non-value-adding work	Total employee involvement and aggressive waste elimination promote speedier operations and eradicate inventories.
8. Lean equipment management	Zero failures, zero defects	Longer equipment life and design improvement reduce cost. Meticulous maintenance and equipment improvement increase quality. Absolute availability and efficiency increase speed.
9. Lean engineering	Zero lost opportunity	Early resolution of design problems with customers and suppliers significantly reduces cost, while improving quality and cycle time.

Figure 1.5   Lean systems purpose.

and JIT, to name a few; the key point is that pursuit of perfection requires discarding current practices that cannot advance performance and inventing new practices that can.

Lean Enterprise Organization

A Lean enterprise organization structure is process centric, focused from end to end on increasing customer value as illustrated in Figure 1.6. Accomplishing this requires definition of company value streams and alignment of organizational functions to those value streams. Value streams are defined as products that are manufactured utilizing at least 70% of same manufacturing processes. Most product-based organizations form their business units around product groups or families but operate in vertical organizational silos. This creates a structural impediment to maximizing an organization's efficient and effective contributions to continuously increasing customer value. Lean organizations are established in a matrix of functional organizations and value streams. The matrix organization's vertical silos of functional disciplines assign resources to be part of value stream teams responsible for products produced by the value stream. Value streams should be owned by business executives of equal status to functional executives, ensuring clear accountability for both functional disciplines and value stream performance. Value stream-based organizations are critical to ensure that

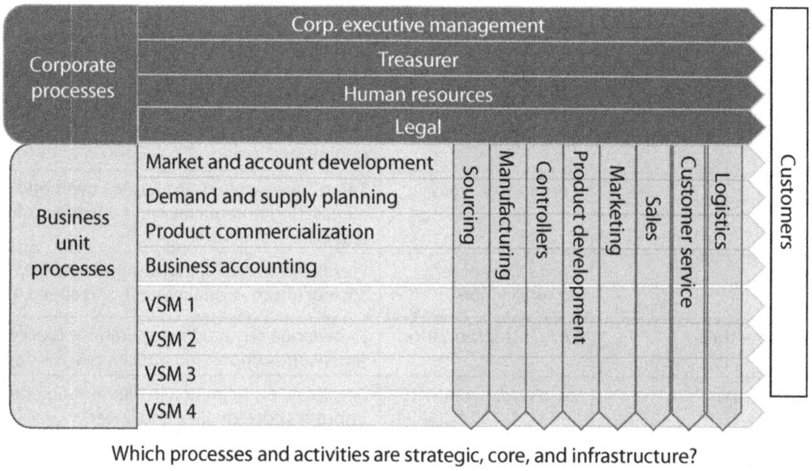

Which processes and activities are strategic, core, and infrastructure?

Figure 1.6 Business process and organizational function matrix.

1. Customer value drives organization priorities.
2. Programs are resourced sufficiently by all functions to achieve their goals.
3. Programs and value stream teams have sufficient leadership to succeed.

This Lean organization matrix creates some additional complexity in managing internal affairs, but this is more than offset by having everyone aligned to increase value for customers. It structurally establishes a balanced focus between customer interests and internal organizational interests.

Lean Application Trends

Although Lean practices have most commonly been applied to supply chain operations, they are equally applicable to all processes and functions in a business. In fact, Taiichi Ohno repeatedly described the TPS as a management system that would work in every type of business. He clearly saw its goal as elimination of waste everywhere it existed in a business. Over the last three decades, greater understanding of Lean along with Taiichi Ohno's and Shigeo Shingo's genius has been slowly discovered. As understanding of Lean practices and principles has grown, they are being applied to all functions and industries.

From Operational to Transactional Processes

Every function from product development to marketing and IT have processes for getting work done. This means value stream mapping system design and PDCA problem solving can be used to continuously improve these processes (see Figure 1.7). A good way to trigger initiation of Lean to technical, sales, marketing, and administrative processes is by starting a dialogue to identify current waste.

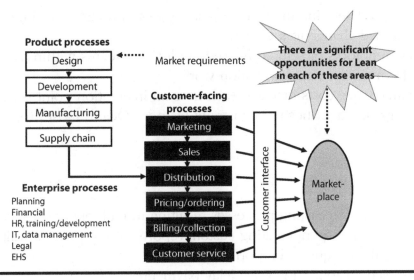

Figure 1.7 Applying Lean to all business functions.

Once participants see there are opportunities, they will want to go to the next step and begin value stream mapping to redesign and improve their processes. Listed below are representative questions that can be used to initiate a waste dialogue in various functions.

Product Design and Development

- Are product development process cycle times too long?
- What causes delays?
- Are launch deadlines missed?
- Why are project milestone dates missed?
- Is the commercialization process inefficient?

IT

- Do managers have access to data they need to run their business—or do they need to fund a project to get the data?
- Are processes defined by the best way to create value for customers or based on implied IT systems processes?
- Can customers get the information they need directly—or do they need to contact a sales/service office to get it?
- Is there one version of the truth for critical enterprise data (such as financials, product sales, customer accounts, etc.)?
- Is there a defined system-of-record for each data set or do users have to hunt for it?

- Is it easy to determine in advance what systems/processes will be impacted by a planned change to a database?
- Are different tools or standards used by different teams within the enterprise to perform similar data integration work?
- Do maintenance costs increase each time an integration point is added?
- Are integration activities the constraint (as per TOC) in most projects?

Finance

- Are inefficient forecasting processes resulting in multiple forecast revisions?
- Is excessive time spent creating custom reports? Do requesters wait too long to receive their requested reports?
- Do disputed customer invoices delay payment?

Human Resources

- Does the hiring process have long cycle times?
- Do associates receive timely feedback and direction?

Customer Service

- Are customers confused about whom to contact for help?
- Do customers have long wait times to get help?
- Do customers get timely responses to complaints?

Marketing

- Are marketing campaigns always effective?
- Are communications with the sales team late, to frequent, to infrequent?
- Is the sales team trained on new product knowledge/expertise to effectively sell the product?
- Are there delays in communicating price increases to the sales force and customers?
- Do pricing errors and price discrepancies exist?

Sales

- How much time are sales associates spending on administrative tasks versus time in the field selling?

- Are sales associates able to follow up on leads in a timely manner?
- Do sales associates get timely responses to their inquiries regarding service issues or are product "tweaks" related to a specific customer?

From Manufacturing to Services Industries

The second biggest trend is from manufacturing to service industries. Many industries such as software development and healthcare have published examples of adopting Lean. Many hospitals and hospital groups across the United States are applying Lean to reduce wait times and increase utilization of doctor's time, equipment, and facilities. Applying Lean is improving quality, reducing cost, and increasing customer satisfaction. Julia Hanna wrote about applying Lean at Wipro, the Indian software development company in her article "Bringing 'Lean' Principles to Service Industries."[7] Two Harvard researchers found that Wipro's initial application of Lean to the development projects increased efficiency by greater than 10% in 80% of the projects. This initial effort grew to 603 Lean projects within 2 years, producing improved productivity and empowered work teams. Their research illustrated five examples of Lean practice application:

1. Using kaizen has altered software development approaches from sequential methods as work moves from one developer to iterative approaches of teams completing software functionality collaboratively.
2. Another example was sharing of mistakes across development teams to learn and apply these experiences to future projects.
3. Wipro used tools like System Complexity Estimator, which compared actual architectures to ideal state architectures, helping teams understand where additional resources would be needed during a project.
4. Value stream mapping was applied to projects, identifying project time waste, leading to increased speed and productivity.
5. Engagement and empowerment energized organizational team members regardless of their level as they saw the bigger picture. This resulted in thousands of software engineers contributing to innovation through problem solving and creating an energized work environment, while increasing productivity and quality. In a white paper called Lean Software Development,[8] Kumar Desai codified Lean software development concepts and practices, making them more easily applied to other organizations. Included in his white paper are identification of software development's eight wastes, nine Lean principles, and how to create a culture of continuous improvement.

Another industry applying Lean to its business is healthcare. David Wessner, president and CEO of Park Nicollet Health Services, St. Louis Park, Minnesota, chronicled their Lean journey in a local newspaper article.[9] Park Nicollet doctors, nurses, and technicians applied standard Lean analysis tools and practices to analyze and radically improve their operations. These included use of stopwatches, spaghetti diagrams, standard work, 5S, level loading, and rapid changeovers, to name a few. The results are transforming Park Nicollet. The first clinic studied was endoscopy, resulting in doubling the number of patients processed each day. This success was replicated in the cancer center, heart center, urgent care, and wound clinic. The use of standard work applied to surgery resulted in 40,000 fewer instruments being used each month; this simplification also meant fewer errors as the right instrument was always available during surgery. In 2004, Park Nicollet completed 85 Rapid Process Improvement Workshops resulting in $7.5 million in savings.

From the Shop Floor to the Enterprise

The third noticeable Lean application trend is "from the shop floor to the enterprise." This subject has been treated previously in this chapter and is the most significant observed trend. Some U.S. businesses have been on their Lean journey for three decades, and only now is Lean enterprise thinking being broadly understood and applied. Lean enterprise application in many respects has been restrained by its great success in manufacturing, creating a perception that Lean is to be applied only to shop floor operations. As previously discussed, this was not the founders' intention, and now we are seeing broader adaptation of Lean as an enterprise management system. The common evolutionary application thread of Lean to organizations is that its broader and deeper application occurs as greater understanding is gained of Lean and the genius of its founders. Practice leads to seeing and understanding more about the organic nature of Lean as a system. Lean is a journey and it is only through practice and study that new doors of understanding and improvement opportunity are discovered and applied. Lean's life cycle is still in the growth phase, which speaks powerfully to future contributions of Lean to radically improving competitiveness of businesses, government services, and even NGOs.

References

1. Shigeo Shingo, *Single Minute Exchange of Dies*, New York: Productivity Press, 1970.
2. W. Edwards Deming, *Out of Crisis*, Cambridge, MA: MIT Press, 1982.
3. Richard Schonberger, *Japanese Manufacturing Techniques*, New York: Free Press, 1982.
4. James Womack, Daniel Jones, and Daniel Roos, *Machine That Changed the World*. New York: Rawson Associates, 1990.
5. Taiichi Ohno, *Toyota Production System*, New York: Productivity Press, 1988.
6. Shigeo Shingo, *A Study of the Toyota Production System*, New York: Productivity Press, 1989.
7. Julia Hanna, Bringing 'Lean' Principles to Service Industries, Harvard Business School, 2007.
8. Dasari Ravi Kumar, Lean Software Development. The Project Perfect White Paper Collection, 2005.
9. David K. Wessner, Toyota System Helps Patients, Health Care, *Minneapolis Star and Tribune*, 2005.

Chapter 2

Why Lean Enterprise?

What Will I Learn?

How being a Lean enterprise will improve my company's strategic and operational competitiveness.

What Is a Lean Enterprise?

A Lean enterprise is one that has adopted the Lean Enterprise Operational Management System (LEOMS) and fully implemented it across all enterprise functions (see Figure 2.1). Being a Lean enterprise is required for twenty-first century companies to have the highest possible chance of achieving sustainable competitive advantage over decades, increasing their probability of existence into the twenty-second century. Only through the adoption of LEOMS can an enterprise achieve continuous improvement through its existence because

1. LEOMS's principles and practices can be applied to all business functions and processes.
2. LEOMS's PDCA problem-solving continuous improvement builds a community of scientists as it is designed into everyone's job.
3. LEOMS engenders a culture of management and team member collaboration.
4. LEOMS is designed to maximize daily continuous total operational cost improvement and not just minimization of labor cost.

Supply chain operations	Administration Finance Legal and HR	Sales and marketing	Product and process development	Information technology

Figure 2.1 LEOMS.

All companies have an operational management system, but is it generating excellence in execution and daily continuous improvement from an army of problem-solving scientists? If not, continue and read this book to learn why and how to implement the LEOMS in your company.

Why Should Every Enterprise Adopt the LEOMS?

U.S. Company Sustainability

In their 1994 book *Built to Last*,[1] Charles Collins and Jerry Porras chronicled 18 companies believed to have shown sustainability. Ten years later in 2004, a Fast Company article, *Was 'Built to Last' Built to Last?*,[2] by Jennifer Reingold and Ryan Underwood looked back at how these companies fared over time. They summarized their findings by saying,

> In the years since *Built to Last* was written, almost half of the visionary companies on the *Built to Last's* book have slipped dramatically in performance and reputation, and their future currently seems more blurred than clairvoyant. Consider the fates of Motorola, Ford, Sony, Walt Disney, Boeing, Nordstrom, Merck and more recently Citi-Group. Each has struggled in recent years, and all have faced serious questions about their leadership and strategy. Odds are, few of them would currently meet *Built to Last* 'visionary company criteria', which requires companies to be a premier player in their industry and be widely admired by people in the know. Besides the stragglers, the list included American Express, General Electric, Hewlett-Packard, IBM, Johnson & Johnson, Marriott, Philip Morris (now Altria), Procter & Gamble, 3M, and Wal-Mart— all of which had far outperformed general markets for decades.

While a few of these companies continued their market leadership and others have recovered some of their prior greatness, long-term sustained leadership performance has eluded all but a handful of companies. No company is guaranteed endless businesses success. The challenge is how to create continuous renewal, ensuring sustainable excellence. Many factors go into competitive advantage, and it is extremely difficult to identify companies that are truly sustainable now or

> "A total management system is needed that develops human ability to its fullest capacity and fruitfulness, utilizes facilities and machines well, and eliminates waste. This system will work for any type of business."
>
> **Taiichi Ohno**
> *Lean Enterprise Operational Management System TPS*

will be 100 years in the future. "If Collins is to be faulted," says James O'Toole, research professor at the Center for Effective Organizations at USC's Marshall School of Business, "it is that he ignores Aristotle's advice not to try to scientifically measure those things that don't lend themselves to quantification."[2] While no one has a predictive model, we can learn from these empirical examples of companies in very competitive markets that have decades of proven leadership success. One observable truth in Robert Waterman's 1987 book *The Renewal Factor*[3] is that valid, sustainable market leadership requires constant attention and renewing. The intention of business strategy is to build and sustain competitive advantage as a market leader. No company "has it made" forever; every dimension of a business must be continually renewed to strengthen its value proposition, strategic processes, and differentiating competencies, so customers continue choosing their products or services.

Becoming a Lean Enterprise Will Enhance Strategic and Operational Competitiveness

The LEOMS will not resolve business model strategic disadvantages, but adopting the LEOMS will optimize potential value of every business model where it is applied vigorously over time. The LEOMS continuously improves a company's value proposition, thus providing the greatest assurance of sustainable competitive advantage. The primary mission of a public company is to provide an attractive return to its shareholders. Companies that have opportunities for growth and better-than-average returns are likely to have price to earnings market valuations higher than market averages. Adopting the LEOMS as their enterprise operational management system creates a continuous improvement process, driving out waste, reducing resource consumption, and creating superior operational capability. This results in a systemic capability to sustain operational competitive advantage by

1. Improving a company's EBITDA (see Figure 2.2)
2. Enabling sales growth
3. Reducing COGS, reducing SG&A cost
4. Substantially reducing capital for new or expanded facilities and equipment
5. Reducing working capital requirements for inventory and accounts receivables
6. Greatly improving long-term return on invested capital
7. Even reducing property tax cost per unit as capacity and sales value per square foot of facilities are greatly increased

How does Lean achieve all these named improvements in operational and financial performance? It results from Lean's capability to deliver superior operational performance built upon Taiichi Ohno's Toyota Production System[3] thinking, values, principles, practices, and tools that make up the LEOMS. The first area is LEOMS's contribution to improving sales. Taiichi Ohno addressed this

The primary mission of a public company is to create shareholder value.
How does Lean create shareholder value? ☑

Increasing net Ebitda, but how?
• Increase profitable sales ☑
• Lower COGS ☑
• Lower SG&A ☑
• Lower taxes ☑

Reducing working capital, but how?
• Lower capital expenditures ☑
• Higher inventory turns ☑
• Lower accounts receivables ☑
• Increase accounts payable days as a % of cash to cash cycle time days ☑

Figure 2.2 LEOMS enables shareholder value improvement.

in describing LEOMS's capability to produce a wide variety of products while achieving service and cost performance superior to conventional batch-based production systems. He directly addressed this apparent conundrum through defining single piece flow as a core principle of LEOMS's foundation. Applying the LEOMS is not just a tactical tool set, but a true transformational process, changing the way organizations think and act. One example is LEOMS's approach to waste reduction. In a typical non-Lean organization, waste has no specific definition and is therefore only obvious when large wastes occur. The organization has technical and management personnel responsible for cost and waste reduction, so they only work on the "big stuff." Focusing on obvious big waste normally happens when some process goes out of control due to a special cause and thus waste reduction is a very reactive improvement process. These organizational behaviors all start simply because they do not have a "waste language," a language to identify wastes in a way permitting their entire organization to see it. LEOMS's eight (see Figure 2.3) wastes define waste with clear definitions that can be observed by anyone trained in its definitions. LEOMS's transformational power is evident through turning an organization's conversation about waste from some amorphous cost to something every team member can learn to identify, determine its root cause(s), and implement countermeasures to

"The Toyota Production Systems evolved out of need. Certain marketplace restrictions required production of small quantities of many varieties under conditions of low demand, a fate the Japanese automobile industry had faced in the postwar period. The Toyota Production System evolved out of need."

Taiichi Ohno
Toyota Production System, p. xiii, Productivity Press, 1988

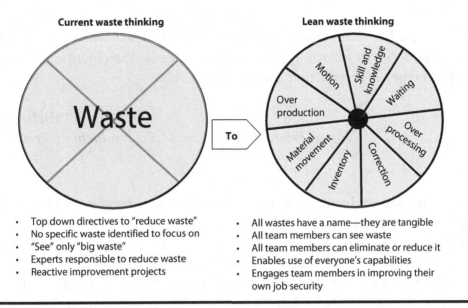

Current waste thinking

Waste

- Top down directives to "reduce waste"
- No specific waste identified to focus on
- "See" only "big waste"
- Experts responsible to reduce waste
- Reactive improvement projects

To

Lean waste thinking

Over production · Motion · Skill and knowledge · Waiting · Over processing · Correction · Inventory · Material movement

- All wastes have a name—they are tangible
- All team members can see waste
- All team members can eliminate or reduce it
- Enables use of everyone's capabilities
- Engages team members in improving their own job security

Figure 2.3   Organizational mindset shift.

eliminate or control it. It turns endless non-value-adding discussions into actionable improvement opportunities. All team members get engaged in waste reduction, improving business performance and their own job security. LEOMS's eight wastes are truly transformational to any organization embracing it by committing to elimination of waste and continually striving for perfection.

Using an economic value-added model (see Figure 2.4), one can understand how LEOMS achieves contributions to reduced cost, increased cash, and market growth. Examining the model, it is possible to trace direct linkages of operational model performance causes (left column of Figure 2.4) to P&L and Balance Sheet performance drivers. These performance drivers affect P&L changes and Balance Sheet elements, which make up Economic Value Added (right-side columns of Figure 2.4). How does the LEOMS improve each of these metrics to levels of excellence achieved by no other existing operational model? How does the LEOMS build capability to produce small lots with lower cost and reduced cycle times? The LEOMS's operational system achieves improved service, lower inventories, and consistently short cycle times in a high variety manufacturing, distribution, and service business environments by applying practices such as flow, level pull, standardized work along with

"All we are doing is looking a the time line" he said, "from the moment the customer gives us an order to the point when we collect the cash. And we are reducing that time line by removing the non-value-added waste."

Taiichi Ohno
Toyota Production System, p. ix, Productivity Press, 1988

"Therefore, all considerations and improvement ideas, when boiled down, must be tied to cost reduction. Saying this in reverse, the criterion for all decision is whether cost reduction can be achieved."[4]

Taiichi Ohno
Toyota Production System, p. 53, Productivity Press, 1988

1. Single minute exchange of dies (SMED) to reduce changeover barriers facilitating efficient small-lot manufacturing
2. Work cell design to ensure consistent production cycle times with maximization of associate value-added time during a shift
3. Level scheduling methods to effectively supply product on time with enhanced productivity by having an even flow through operations
4. Pacemaker scheduling to schedule a single operation that synchronizes continuous flow of remaining operations and suppliers through pull systems and supermarkets
5. Application of error proofing to eliminate product, order fulfillment, and shipping errors that have a negative effect on productivity and material cost

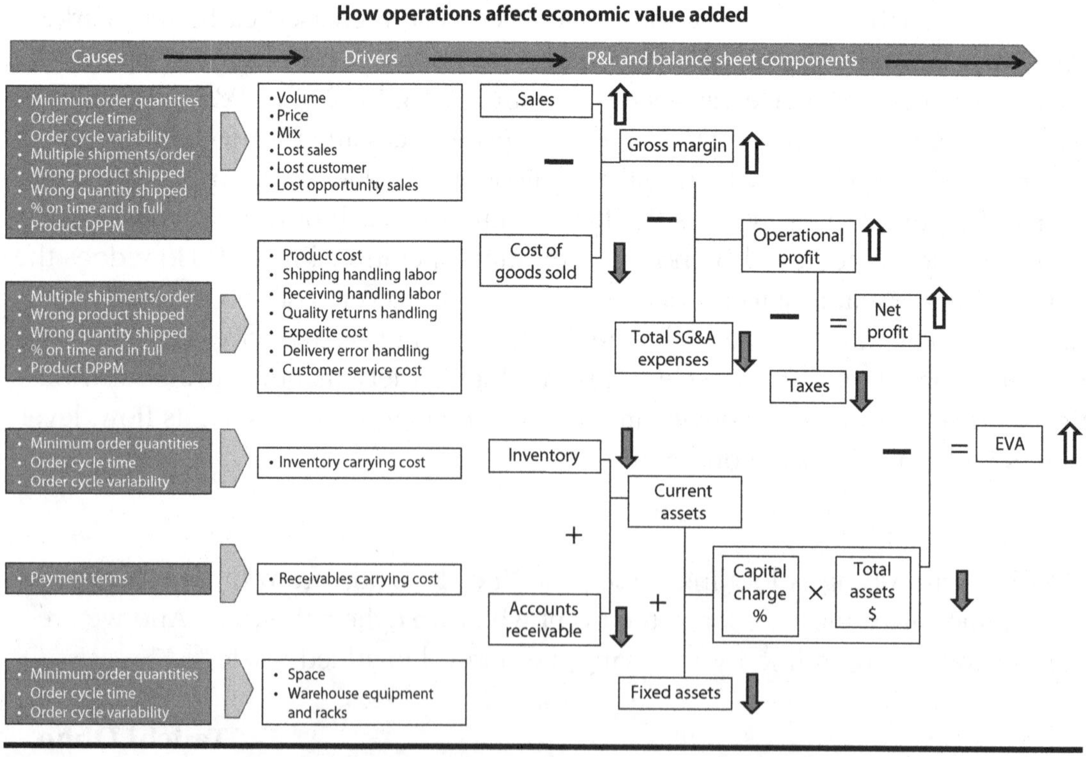

Figure 2.4   Economic value-added model.

"At Toyota, we go one step further and try to extract improvement from excess capacity. This is because, with greater production capacity, we don't need to fear new cost."

Taiichi Ohno
Toyota Production System, p. 57, Productivity Press, 1988

Applying these practices has a positive effect on growth as a result of

- Significantly faster and more reliable delivery of orders on time and in full
- Smaller customer minimum order size requirements
- Reduced quality issues
- Lower operational cost

The LEOMS breaks a historically accepted paradigm, which states that small lot production is mutually exclusive with cost leadership. LEOMS also makes a very significant contribution to supporting growth by revealing hidden capacity to support growth very cost effectively, as only variable costs need to be added to produce additional product volume. While it is true that increasing available space and equipment capacity only has an immediate effect on cost if there is volume to utilize them, it is also true that it is a countermeasure against future company growth cost.

LEOMS Focuses Organizational Resources and Processes

LEOMS is often thought of as only having relevance to product and service operational processes such as supply chain operational cost, working capital, and capital expenditures. LEOMS applies to all company processes and functions (see Figure 2.5)

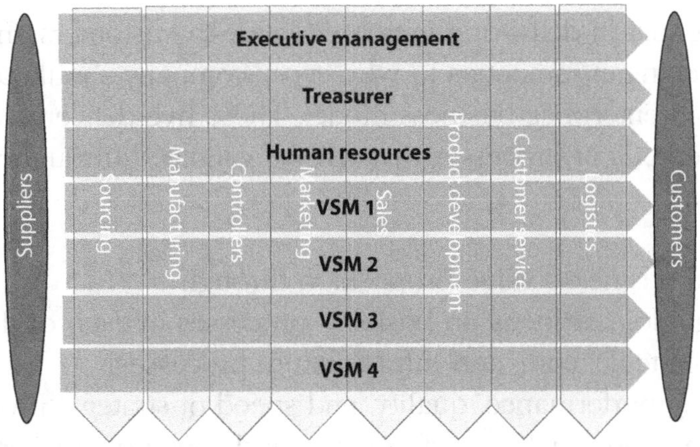

Figure 2.5   Lean enterprise organization.

resulting in improved organizational alignment, shorter process cycle times, increased process reliability, and reduced cost through elimination of waste. This includes reducing cost of SG&A processes and functions. Lean enterprise creates a horizontal organization of product and service operations from suppliers to customers, aligning operational value creation with customer expectations within value streams. Value streams are defined as groups of products that share processes (usually with at least 70% common processes). Value streams may serve more than one supply chain as supply chains are defined as a logical group of products serving a defined set of customers. For example, two supply chains, one serving vehicle manufacturers and the other serving appliance manufacturers, may well be sourced from a single value stream. This simply means the most demanding supply chain requirements, regardless of the end market, are applied to the value stream supplying these different market segments. Lean improvement efforts must be focused on strategic creation value of the entire organization. For example, marketing focuses on building strong brands and loyal customers who are repeat buyers and often tell friends their opinion of brands they have used. Lean efforts applied to improve processes enable marketing to increase brand strength and market share in a cost-effective way. One only has to remember that Lean's fundamental purpose is to create increased value for customers. In research and development processes, Lean should be applied to improving speed and quality of understanding customer habits, use of current products, and unmet needs, enabling faster design and development of new innovative products improvement of existing products. Lean tools and practices are applicable to all transactional SG&A processes, eliminating waste, improving quality, and decreasing cycle times. The LEOMS is a complete operational management system for all enterprises; it is simply a matter of intelligently applying LEOMS's practices and tools to deliver increasing enterprise strategic value from every function and process.

Five Demonstrable Reasons Why LEOMS Is the Best Operational Management System

LEOMS has no equal in delivering increased value to customers and investors, making a compelling argument as to why every company should aggressively embrace the Lean enterprise business model. These five deliverables are supported by 30 methods or processes that will be validated throughout the remainder of this book.

1. Delivers the business's value proposition through
   a. Creating a logical focus on business processes in each of the three segments (strategic, core, and infrastructure processes)
   b. Improving performance, quality, and speed of strategic processes
   c. Improving cost, quality, performance, and speed of core processes, prioritizing those with the greatest contribution to strategic processes
   d. Improving cost, quality, and speed of infrastructure processes

    e. Explicitly identifying differentiating organizational competencies, facilitating assessment, improvement, and investment to increase strategic competitive advantage

    f. Defining the value proposition explicitly through identification of differentiated competencies and strategic processes

    g. Linking business plan-driven targets and goals to a value proposition and value discipline (Hoshin Kanri)

    h. Prioritizing organizational focus on the vital few targets (Hoshin Kanri)

    i. Ensuring complete cross-function goal and priority alignment (top-down and bottoms-up; Hoshin Kanri)

    j. Applying a planning process that builds balanced goals; internal–customer, strategic–operational, long term–short term (Hoshin Kanri)

2. Builds added value from suppliers to customers by

    a. Setting end-to-end supply chain goals (supplier to customer value stream targets)

    b. Including all organizational management levels in operational planning (Hoshin Kanri)

    c. Expecting all supply chain participants to contribute to improvement goals (LEOMS's problem solving)

3. Implements best practices and encourages benchmarking because the LEOMS

    a. Is a complete operational system (Lean Enterprise Operational Management System)

    b. Contains leading practices that are applied as a part of LEOMS's implementation (LEOMS improvement practices)

    c. Drives for perfection and looks for better practices (Lean principles)

4. Creates customer value as a result of

    a. Building strategic capabilities through prioritizing the focus on strategic processes and differentiating competencies (Lean enterprise process segmentation)

    b. Scheduling operational planning and review cycles supporting timely adaptation to market changes (LEOMS reviews)

    c. Flexibly designing operations to quickly adapt to market changes (Hoshin Kanri regularly scheduled reviews and daily continuous improvement)

    d. LEOMS's ability to automatically adopt to volume and mix variability with minimum cost penalty (Lean value streams)

5. Drives continuous improvement forever as it

    a. Holds all levels of leadership accountable (Lean operational reviews)

    b. Establishes annual improvement targets (Hoshin Kanri)

    c. Constantly increases customer and shareholder value (continuous improvement process)

    d. Provides a common language and tools (Lean operational management system model, tools, and processes)

    e. Keeps improvement gains (standard work and tiered process audits)

f. Applies scientific problem solving (Lean PDCA)
g. Creates and drives toward an end state vision (Lean value proposition and Hoshin Kanri)
h. Utilizes an end-to-end business process assessment (value stream mapping)
i. Builds a culture where associates do and improve their job (Lean associate role)
j. Engages the entire organization (only operational associates add value)

The remaining chapters focus on LEOMS application implementation and will be explained in a logical step-by-step method. In the process, how the five attributes discussed above can become a reality of any company that is willing to make the commitment to becoming a Lean enterprise will be shown.

References

1. Charles Collins and Jerry Porras, *Built to Last*, New York: Harper Business, 1994.
2. Jennifer Reingold and Ryan Underwood, *Was 'Built To Last' Built To Last?* New York: Fast Company, 2004.
3. Robert Waterman, *The Renewal Factor*, New Providence, NJ: Bantam, 1987.
4. Taiichi Ohno, *Toyota Production System*, New York: Productivity Press, 1988.

Chapter 3

Business Strategy-Driven Lean Enterprise

What Will I Learn?

How to adapt and strengthen your company's business model by adopting LEOMS as a key element.

Strategy-Driven Lean System

The purpose of a business is to create value for customers and a financial return for investors. This requires a customer-focused strategy in defined market segments and products or service offerings they will purchase because they are perceived as providing the best value. In return, companies providing these products or services typically enjoy a significant market share, higher growth, and profitability than their competitors. Sustaining a leading market position is only achieved by continuing to improve and/or reinvent attractive offerings at a rate faster than competitors. In order to support continuous reinvention, a company must invest in technology development requiring significant financial resources. In addition, business strategy implementation depends on an operational model designed and operated to effectively achieve strategic goals and operational targets. LEOMS's proven capability to produce exceptional financial results for any given business model validates it as the best operational management system or model known. Purposefully designed by visionary thinking and perfected through practical application, the LEOMS is the best-known solution for meeting challenges as described by Professor Simchi-Levi.

Companywide fully committed LEOMS applications always produce operational improvement, in alignment with business strategy, producing both operational improvements and greater strategic competitive advantage. A Lean enterprise is not about simply applying Lean thinking, principles, and practices to operations to improve cost and reduce inventories, but applying LEOMS's power to continually

> "The new normal is a business world that is complex, uncertain, dynamic, and chaotic. Companies must craft a supply chain strategy that handles these realities."
>
> **David Simchi-Levi**
> *Professor of Engineering Systems at MIT, Operation Rules: Delivering Customer Value through Flexible Operations in High Tech Supply Chains, Harvard Business Review, March 2011*

increase strategic competitive advantage. The LEOMS accomplishes this by continuously

1. Focusing on customer requirements and desires in order to meet/exceed customer expectations
2. Reducing the eight wastes, resulting in lower operating cost and capital requirements, thus providing resources to continually reinvest in technology innovation and new product/service development
3. Engaging the entire organizations in building and applying knowledge every day not only to solve defects and deviations but also to continually improve processes by defining and achieving the "next target level conditions"
4. Driving all aspects of operations toward "perfection target conditions"
5. Building an army of problem-solving scientists who make sustainable improvements every day

How Does the LEOMS Increase Profit?

Earnings per share growth is the primary measure of a company's financial success, and its price-to-earnings ratio is a measure of market confidence in future earnings growth. Earnings growth is a result of profit margin and revenue growth. To be sustainable, revenue growth must be supported by products perceived as value to customers, or it will be vulnerable to a competitor's new product, price decrease, or promotion. So, when discussing growth, it needs to be quality growth, which is valued by customers. This level of delivered customer value defines the barrier competitors have to overcome for customers to change.

> "Therefore, all considerations and improvement ideas, when boiled down, must be tied to cost reduction. Saying this in reverse, the criterion for all decisions is whether cost reduction can be achieved."
>
> **Taiichi Ohno**
> *Toyota Production System, p. 53, Productivity Press, 1988*

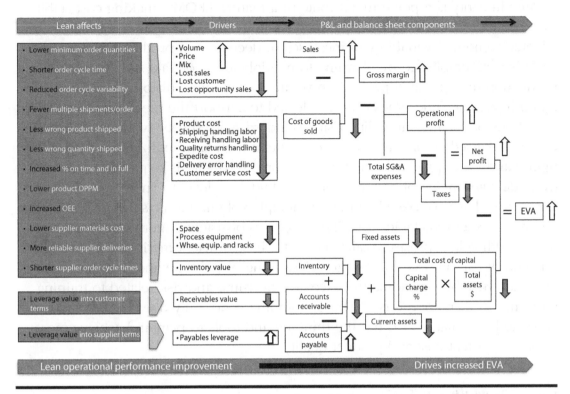

Figure 3.1   The LEOMS increases economic value added.

Reduces Cost of Goods Sold

The LEOMS reduces cost of goods sold (COGS) by having a positive impact on economic value-added "drivers" (see Figure 3.1). By focusing on value stream waste elimination, direct and indirect labor costs, material costs, and available capacity are improved. Starting with Lean Affects (left-hand column), they are linked to EVA Drivers (in the second column from the left). The arrows inside the Driver's text boxes indicate the effect of Lean improvements on the various drivers.

Reduces Labor Costs

Productivity benefits show up immediately in COGS, assuming additional volume can be sold or people who are no longer required can be redeployed to other value-adding activities. Elimination of product quality and shipping failures also contributes to both lower cost and higher operational predictability. In addition to reducing value stream cost, Lean also increases speed and reduces cycle time variability resulting in more predictability of costs and therefore improved credibility of company earnings forecast. In most companies, when volume goes up, labor cost declines as a percentage of sales. This indicates that the system is carrying excess labor; otherwise, labor resource requirements would be linearly related to volume. In a mature LEOMS, direct labor is optimized to have high value-added utilization at the planned levels of production, resulting in labor costs maintaining a linear relationship to volume. Direct labor hours to output volume linearity is a powerful attribute of a mature LEOMS, making cost stability a norm. System flexibility is planned so crew sizes and work assignments are adapted to normal variability of increases or decreases in production demand.

This is accomplished by planning to multiple levels of Takt time (available production time/planned production volume). Examining historical variation in monthly volume, Takt times can be selected to support the mean or two standard deviations of volume variability. Using standard work sheets, work cell operations can be balanced among associates to maintain high labor utilization at selected higher and lower production levels. Workforce flexibility is required to cost effectively take advantage of moving from one Takt time level to another in as short a time interval as possible. Planning for multiple volume levels is a significant effort and requires a predictable variability and/or a consistent pattern of growth that may be difficult to precisely calculate, but it has sufficient merit that it should be pursued. It can be used for workforce planning of full-time and part-time front-line team members avoiding the excessive overtime and risk related to training part-time workers in a "crisis." The bottom line is not only does LEOMS reduce absolute labor cost; it also has the thinking and tools to plan and achieve more stable, consistent cost and product quality.

Reduces Materials Cost

Materials cost is reduced through elimination of scrap, inherent waste, and process waste:

1. Scrap is typically input materials that are defective and therefore never processed but thrown in the dumpster.
2. Inherent wastes are often weed and/or trim wastes designed into many processes and their tooling.
3. Process waste is the percent of input materials converted to good output.

In many companies, knowledge and data related to these three wastes are unknown as either systems do not capture all waste, it is not reported properly, or reporting is only done on wastes that are "controllable" by operations. The "dumpster dive" often is a gold mine of opportunities for these three forms of material waste. Emptying the dumpster on the floor and then sorting scrap by type of waste is a simple and effective way to start working on material waste elimination and engaging the workforce in PDCA. If your company does not have detailed reporting, try dumpster diving; just dump out the dumpster and identify all the waste.

Reduces Future Fixed Asset Cost

While existing fixed assets are sunk cost, it is possible to reduce future capital cost requirements for supporting business growth. LEOMS's thinking values

> "At Toyota, we go one step further and try to extract improvement from excess capacity. This is because, with greater production capacity, we don't need to fear new cost."
>
> **Taiichi Ohno**
> *Toyota Production System, p. 57, Productivity Press, 1988*

creating machine capacity as a hedge against increased cost of growth by using available machine capacity so only variable cost needs to be added to produce more volume. Measuring and improving operational efficiency and effectiveness (OEE) of key process operations increases capacity. OEE percent is calculated as follows: average good product cycles per scheduled shift hour divided by designed cycles per scheduled shift hour. It measures effective use of the maximum potential process cycles during scheduled shift hours. It is a measure of a process's value-added time. Its complement is waste or opportunities to gain additional effective process cycles and thus increased capacity. In the long term, this results in fewer required pieces of equipment, less floor space cost, lower operational maintenance cost, and lower equipment replacement capital cost.

Reduces Working Capital Cost

Improving process cycle times and consistency of cycle times is another primary focus of LEOMS. This will result in less inventory due to shorter lead times, less safety stock requirements, and less cycle time variability. In addition, shorter and more consistent cycle times increase product delivery reliability and reduce customer costs of ownership enabling an opportunity to negotiate more favorable payment terms with customers. On the payables side, investing in partnerships with suppliers and adding value by sharing Lean knowledge through joint projects in supplier's facilities, opens the door to shared cost reduction benefits, and/or negotiating more favorable payment terms.

Enables Increased Top Line Sales and Reduces SG&A Cost

Lean enterprise transformation reaches beyond supply chain operational improvement and leads to improvement in growth and earnings metrics (Figure 3.2). For example, credibility and confidence are maintained based on how a company meets its product and service commitments.

Customers' view of the ease of doing business with a supplier is influenced by their experiences. Lean enterprise focuses on customer value and elimination of every activity not adding customer value, from all company processes, not just supply chain operations. A customer's total experience with its supplier's products and service is considered in making their choice of preferred suppliers.

Earnings	Earnings per share CAGR
Sales	Revenue CAGR
	Market share
Income	Net income CAGR
	EBITDA CAGR
	SG&A %

Figure 3.2 Sales, income, and earnings metrics.

Enables Increased Brand Value

In addition, customers' experiences with a company's people, products, and services are the substance of brand reputation. No amount of advertising and promotion will compensate for customer experiences that conflict with a company's brand messages. Lean enterprise offers the promise of systematically pursuing perfect alignment of a company's brand messages and customer experiences.

Sales Growth

There are three primary possibilities for sales growth: first is market-driven growth or just growing along with the market; second is taking share from competitors; and third is entering new markets or product categories. Lean cannot compensate for a lack of sales, marketing and customer service competencies, nor insufficient research and development investment, but what it can do is greatly improve effectiveness of existing resources and competencies. It can increase existing skilled people's value through elimination of non-value-added activity in their everyday processes. This elimination of waste results in faster sales cycle times from qualifying leads to closing the sale. In product development, it means becoming faster to market whether it is with a new or improved innovative product or fast following a competitor.

SG&A Cost Reduction

SG&A cost for most companies is their second largest cost next to COGS. In some industries, business models with low factory cost have high research and development, sales, customer support, and regulatory costs resulting in SG&A being their highest cost. So, it is not just operations that must be examined to maximize customer value; all SG&A functions and processes must be included in a Lean enterprise transformation to build and sustain competitive advantage.

What Are the Requirements for Sustainable Market Leadership?

In their book, *Discipline of Market Leaders*,[1] Michael Treacy and Fred Wiersema state that market leaders focus on one of three market disciplines as the source of their competitive advantage: (1) *operational excellence*, (2) *product leadership*, and (3) *customer intimacy*. They also defined four rules market leaders follow and what LEOMS contribute to the success of each of them.

Rule 1: Provide the best offering in the marketplace by excelling in one of the market disciplines: (1) operational excellence, (2) product leadership, and (3) customer intimacy. Regardless of the market discipline chosen, Lean enterprise is the best enterprise business model and operational management system to achieve sustainable chosen market discipline success through continuous improvement.

Rule 2: Maintain threshold standards on other dimensions. Maintaining threshold performance means you must have industry parity on all dimensions of competitive performance to avoid being at a disadvantage versus market competitors. The speed of global markets is continuing to raise the bar on minimum threshold levels, below which a company is competitively disadvantaged.

Rule 3: Build a well-tuned operating model dedicated to delivering unmatched value. Rule 3 is the real power and LEOMS's differentiation. It means having well-defined differentiating competencies, strategic business processes capability, and defined performance attributes maintained at the market's highest levels by rigorous application of Lean's organization-wide continuous improvement approach.

Rule 4: Dominate your market by improving year after year—LEOMS's proven continuous improvement process has decades of sustained gains demonstrated by leading companies. There are several companies that have created markets and have lost their market dominance because they did not continually reinvent themselves in the markets they created.

These rules give context and relevance to the treatment of all nine elements of the Lean enterprise business model (see Figure 3.4); their implementation is critical to building a Lean enterprise. They should be adopted by companies embarking on the LEOMS implementation. In most companies, no one can succinctly define and explain their business model and operational system, even though it has to exist for them to operate. Companies must have a winning business strategy and operational management system excellence sufficient to achieve top line growth and improved operating income faster than competitors.

Lean Enterprise

Lean enterprise (see Figure 3.3), is built on LEOMS's complete operational management system that includes all enterprise business function processes (see Figure 3.4).

Lean Enterprise Business Model

It is a strategic decision to transform a business into a Lean enterprise because the LEOMS is an operational management system designed to enable sustainable market leadership when driven by a winning business strategy (see Figure 3.4). Success with Lean needs to be viewed as a strategic organizational change as it will only be effective and durable if it truly becomes a part of a company's

Lean enterprise operational management system				
Supply chain operations	Administration Finance Legal and HR	Sales and marketing	Product and process development	Information technology

Figure 3.3   Lean enterprise.

Strategic components

Business vision, mission, and values

Business culture: customer first-team member development-process focus-continuous improvement

Business strategy must define superior differentiation versus market competitors:
• Market discipline: product leadership, customer intimacy, or operational excellence
• Differentiating competencies: competencies create the desired customer choice attributes
• Value proposition: i.e., niche applications, total solutions, low cost, etc.
• Strategies: actions to be the customers' preferred choice

Operational model components

Operating models must be designed and operated to achieve and sustain competitive advantage through excellence in the chosen value discipline and proposition:

Business plan: three-year operational plan (year 1 in complete detail)

Vital few metrics: critical few vital metrics aligned with key strategic goals and targets

Business processes:
• Processes to create, supply, and enable value for customers – strategic, core, and infrastructure
• Differentiating competencies – distinctive skill sets and activities within processes producing product and/service attributes that attract customers to purchase.

Business structure:

Go to market model: OEM-end user/retail/channel partner

Organizational management elements: org. structure, policies, rewards, financial systems

Lean enterprise operational management system: strategic plan deployment, plan and execution management; material flow, operational process design, and continuous improvement processes

Figure 3.4   Lean enterprise business model.

culture. The Lean enterprise business model is illustrated with two sets of components: strategic components and operational model components.

Strategic Components

The strategic components define what we want to accomplish and who we are as an organization; it satisfies Rule 1: provide the best offering in one of the market disciplines—operational excellence, product leadership, or customer intimacy.

Business vision, mission, and values: Lean enterprise must be an integral part of a company's vision, mission, and values in order for it to be successful in achieving its goals. It requires values that respect every job and person in the organization. In fact, being a Lean enterprise is often discussed in the context of the "inverted organization chart" with the frontline team members who perform operational processes at the top supported by increasingly higher organizational levels and executives at the bottom.

Business culture: The decision to adopt the LEOMS is a strategic decision because it has widespread company culture implications, as being successful means Lean will alter behaviors and thinking of every individual in the organization. Culture guides an organization's behaviors, norms, and practices: the unwritten rules defining such things as how things really work, what is really expected, and how to succeed. Introducing Lean thinking, processes, and tools has enormous implications related to company culture, as existing company culture is likely to be a significant Lean transformation risk factor. In order to eliminate this risk, company culture must be assessed and, if necessary, modified. Current company culture also represents the biggest organizational opportunity for significant companywide sustainable improvement through internalization of Lean thinking and practices. Company culture assessment and change will be discussed in more detail in Chapter 4.

Business strategy. This is not a book about business strategy or strategic planning, but effective implementation of Lean must be completely aligned with business strategy and viewed as the primary engine for making strategic customer experience intentions a reality in the lives of customers. Lean implemented without alignment to business strategy may produce some short-term results but will never achieve its potential in contributing to long-term market leadership and above market average financial results sustainability.

Operational Model Components

Operational model components define how strategic plan targets will be accomplished to satisfy discipline of market leaders' rules 2, 3, and 4:

Rule 2: building a well-tuned operating model to deliver unmatched value
Rule 3: maintain threshold performance as a minimum on all performance metrics
Rule 4: dominate your market by improving year after year

Every company has an operational system, and whether designed or evolved, there is an existing model. To be effective, an operational system's performance and processes must be constantly aligned and measured to deliver a company's value proposition to customers, consistently and efficiently. It must be managed in rigorous detail as customer experiences are determined by perfect operational detail execution. Effective operational models must be aligned to a well-defined value discipline and value proposition that will make a company better operationally and strategically to ensure long-term sustainability.

Lean enterprise has significant implications related to each of its six operational model components (see Figure 3.4):

1. *Business plan*: The LEOMS provides an excellent operational planning process called Hoshin Kanri, to be discussed in Chapter 5; but suffice it to say that it is an extremely effective process for assuring that operational plans are focused to achieve strategic plan improvement targets.

2. *Vital few metrics*: A key output of Hoshin Kanri are the vital few metrics, providing strategic True North, aligned with a company's value proposition and strategies to achieve 1- and 3-year strategic, operational, and financial targets. These targets drive LEOMS's redesign and operation improvement, keeping organizations on track by having a laser beam focus on increasing their competitive advantage.

3. *Business processes*: Business processes are the operational model's third element; they are sets of activities done throughout a company's functional organization to get work done. Business processes are classified into three categories: strategic, core, and infrastructure.

   a. Strategic processes create differentiation in a company's delivery of unsurpassed customer value to their customers as promised by their value proposition.

   b. Core processes complement strategic processes in executing operational activities that get the work done in support of the strategic processes to deliver the value proposition.

   c. Infrastructure processes are typically back office processes that are not directly part of products or services affecting customers but essential for effective core and strategic process execution.

   Strategic planning identifies necessary business process improvement targets as one of its outputs that become key inputs into Hoshin Kanri, assuring that proper priority, resources, and action plans are in place to achieve required strategic plan performance targets.

4. *Business structure and organizational design*: Businesses are most commonly organized around products, technologies, or markets. In many cases, when there is sufficient business scale, they are hybrids with highest organization units focused on markets or technologies and the second level of

units organized by market segments or product families. The importance of understanding this is related to determining logical ways to adapt current business and organizational structures to Lean's end-to-end value stream management approach (Figure 3.5). Value streams are a company's supply chain information and material flow, extending from suppliers to customers. While this represents another dimension of organizational complexity, it adds tremendous value by creating focus on relentlessly adding customer value. At some point in a Lean enterprise transformation journey, every organization must address realigning its operational management and support resources around its value streams.

5. *Go to market models*: Often several value streams will provide product to supply chains aligned with the companies' go to market model(s), OEM, end user, retail, or channel partner requiring specific supply chain designs to meet the varied service requirements of these four go to market models.

6. *Management systems*: An organization's management systems, such as policies, metrics, performance appraisals and rewards, financial processes, and their supporting IT systems. Lean enterprises are hardwired to deliver customer value as their value streams reach from their suppliers through to end customers.

Lean enterprises are designed and operated based on Lean culture assumptions and principles that transform companies only when there is internal integrity and consistency of leaders' behaviors aligned with Lean people leadership expectations. Lean organizations have a maniacal daily focus on continuously improving every day and thereby increasing value to customers in an endless pursuit of perfection.

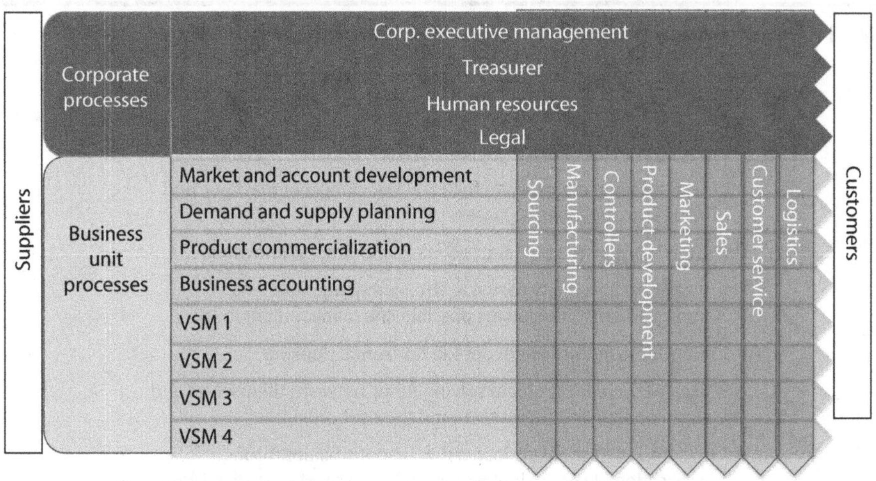

Business process and organizational function matrix

Figure 3.5   Lean enterprise organization.

Lean Enterprise Business Model Implementation Framework—A Roadmap

The Lean enterprise business model implementation framework is provided to assist in integrating LEOMS as the company's business model (see Figure 3.6).

1. *Vision, values, and culture*: LEOMS implementation success requires an aligned and enabling company vision, values, and culture that will inspire, develop, and nurture an organization's daily execution of LEOMS's seven people leadership expectations (see Figure 3.7; they are discussed in more

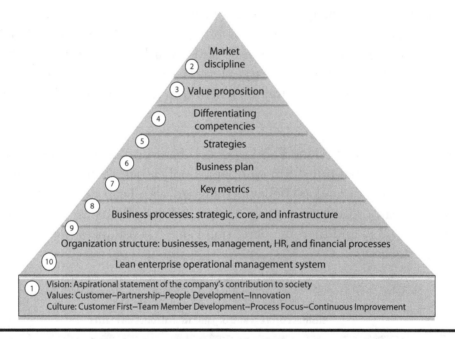

Figure 3.6   Lean enterprise business model implementation framework.

1. Engages people at all organizational levels

2. Teaches team members to focus on the work, material flow, and value stream to see the waste

3. Gives team members deep technical and process knowledge

4. Pushes responsibility for value stream management and improvement to the lowest possible line management level

5. Introduces metrics to encourage horizontal thinking

6. Creates frequent problem solving loops between themselves and their superior and themselves and their subordinates

7. Accomplishes these six leadership responsibilities through application of policy deployment, A3 analysis, standardized work with standardized management and kaizen.

Figure 3.7   LEOMS's leader people leadership expectations.

detail in Chapter 6). This is the biggest and most important challenge that must be successfully overcome for long-term LEOMS success.

2. *How to define the market discipline?* Defining a market/value discipline: product leadership, customer intimacy, or operational excellence. The question needing an answer is, what are the dominant reasons customers repeatedly purchase products and/or services from your company? See Figure 3.8, as an example of defining a value discipline; in this case, product leadership was selected to comply with rule 1: provide the best marketplace offering by excelling in a specific value discipline.

3. *Value proposition.* Understanding value propositions starts with a rigorous assessment of product and service attributes that explain why customers choose a company as their supplier. Every company has a value proposition, but

   a. Unfortunately, for many it is not explicitly defined and understood by all company team members. This is a long-term survival risk and renders short-term operational improvement efforts ineffectual in impacting increased enterprise strategic competitiveness.

   b. Another frequent situation is companies start by making something that was a success at least for a time without considering what their real

Value discipline – Product leadership
Value proposition – Product innovation

Operation model

Business structure
> Industry/market facing sales, marketing, and technical service organizations
> Product focused business units and teams
> Strong product marketing, sales, applications, technical service, and product and process development organizations

Management systems
> Market and product P&L structure

Culture
> New technology and product heroes
> Product P&L gross margin drives decisions
> Technology knowledge sharing and mentoring
> Growth through new products and expanded applications of existing products and technologies

Strategic processes

> Process design, development, and commercialization
> Intellectual property management process
> Materials technology research and development
> Technology and application intellectual property
> Global technology scanning, evaluation, and acquisition
> Manufacturing process design, proprietary equipment design, and operation management

Differentiating activities

> Relationships with key customer technical management and knowledge of their products and services
> Knowledge of customer design, development and production processes, and key personnel
> Participation in customers' product development process
> Relationships with customer operations management teams

Priority metrics

> Product category penetration
> Percent of sales from new products
> Sales growth
> Order delivered on time and in full
> Product quality (DPPM)
> Upside flexibility
> Cost of goods sold percent
> Return on capital employed

Figure 3.8 Product leadership value discipline.

market discipline is and what competencies they needed to focus on to sustain their success. The market is full of short-life one-product wonders.

It is possible for a company to change its value proposition, but this is a very difficult and a multiyear process and few have been successful. One example of success is IBM as they transformed themselves from a computer hardware company to an information services provider with a focus on the strategic use of analytics and consulting services. This type of change takes courageous leadership with great vision and leadership skills. IBM is a rare case, as most businesses do not succeed in making a strategic transformation into another business. IBM has the advantages of enormous research and development capabilities, a large patent portfolio, and long history of working with clients' information systems and data. Specific value propositions fit within the larger framework of a company's overall market discipline. The question is whether or not their new business model is sustainable long term or will require another reinvention, but one has to give them credit as few companies have been able to achieve a true reinvention and become successful again. While a value proposition within a discipline is required to establish differentiation from competitors, only with continuous operational improvement can a value proposition maintain competitiveness and provide investors with maximum potential financial returns.

4. *Differentiating competencies.* Value propositions are enabled by a set of differentiating competencies and strategic business processes creating differentiation of a company's market discipline. Differentiating competencies are processes and activities that create and deliver a company's value proposition to their customers, and they are the source of their value proposition differentiation. Competencies are the internal processes that differentiate a company's product or service. While these selected competencies are true sources of competitive advantage, they are delivered to customers through business processes made up of many steps and many activities. Some of the processes are strategic, some core operational, and some infrastructure, the processes supporting and enabling strategic and operational processes. For example, suppose that customers buy a product because of value delivered from its unique design features; this would infer excellent insight into customer applications and product design, as this is what customers observe from their experience. (Steve Jobs and Apple Computer was the standard in these competencies.) They consistently designed products that met customers' unarticulated needs and desires. These competencies are described as strategic competencies, intended for customers to experience and be delighted with products or services they purchase. To ensure customers do experience this delight, many process steps must be completed efficiently and effectively or they detract from a customer's good experience. These processes such as processing orders are not strategic but operational, and processing orders requires routine work activities such as credit checks that are infrastructure processes. Think of infrastructure

processes as those that go on in the back room and are not part of the daily execution of designing, selling, making, shipping, and billing processes, but they have to be completed without error and timely to ensure that operational and strategic processes do not fail and result in customer dissatisfaction. In the example shown in Figure 3.8, differentiating competencies were focused on being engaged directly with customer organizations in order to build knowledge regarding current applications and future product developments to be able to continually improve current products, and find opportunities to contribute to the success of their customers' next-generation products and services.

5. *Strategies.* This book is not focused on strategic planning, and there are many excellent sources for assistance with strategic planning. Strategic planning has been described as "informed opportunism" as there is not a cookbook formula to follow for creating successful business strategies. The overarching intention of business strategy is to provide direction as to how a business is going to profitably grow with existing and new customers faster than competitors providing attractive returns to investors. Strategic plans are critical to providing clarity and focus to operations regarding improving capabilities and performance. Resources are always limited, so the more focused the direction coming from strategic plans, the more valuable operational improvement initiatives and results will be.

6. *Business plan.* Hoshin Kanri or policy deployment with method is an effective strategic and operational alignment of organizational metrics, programs, and resources that will
   a. Level load consumption of resources to match demand
   b. Engage and align all organizational levels through the catch ball communication process
   c. Align short- and long-term strategic objectives
   d. Establish a rhythm of periodic reviews
   e. Build rigor and discipline via PDCA, creating a closed-loop execution management system

7. *Vital few metrics.* The LEOMS drives top and bottom line results.

Performance Metric Linkage to Customer Value

Market leading businesses win by focusing on customer value-driven metrics and goals defined from Hoshin Kanri annual operational business plans. Business leaders need to stay close to their customers so they understand the "customer's voice," their needs, expectations, and priorities, which must be satisfied to have sustained business success. Use of customer economic value add (EVA) models (Figure 3.9) leads to understanding strategic metric impact on a customer's EVA. A company's few vital metrics or key performance indicators (KPIs) must have a "line of sight" to customer value and expectations. This assures a company's metrics are well enough aligned with those of their customers and are valid predictors

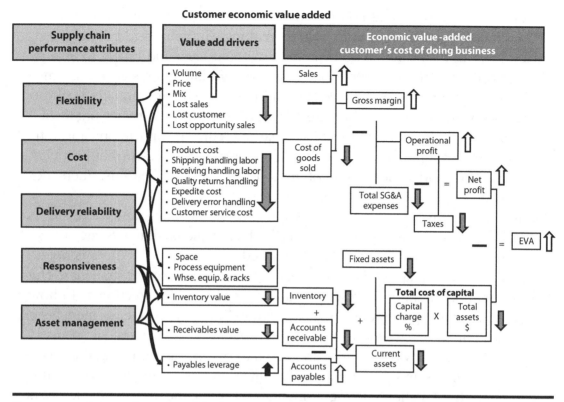

Figure 3.9 S.C. performance drives customer economic value added.

of customer satisfaction and value added. EVA measures enterprise financial health. It is based on a financial principle of measuring how much net profit exceeds the total cost of invested capital. Figure 3.9 illustrates a customer EVA model. EVA (far right side of the chart) results from subtracting total cost of capital from net profit. Net profit is derived by subtracting cost of goods sold, sales, general and administrative cost, and taxes from sales revenue. Total cost of invested capital is a result of multiplying the cost of capital percent by total company assets. To the left of net profit and total cost of capital are components that go into their calculation. On the left-hand side of Figure 3.9 are factors that drive EVA sales revenue, cost, and asset components. Suppliers must focus on how to improve the "value add drivers" in Figure 3.9, representing their key customers' highest priorities. EVA, whether used as an analytical tool or as a discussion framework, will contribute to a deeper understanding of how to drive increasing customer value.

How to Align Operational Performance to Customer Value Priorities

Supply chains can be described by their performance attributes. The inherent performance capability of each of these attributes is determined by a supply chain's design DNA and operational effectiveness. There are five high-level supply chain attributes, shown in Figure 3.9, whose inherent capability are designed into every supply chain.

a. *Delivery reliability*: dependability in getting products to the customer on time
b. *Responsiveness*: the amount of time it takes to fill and ship customer orders
c. *Flexibility*: ability to respond to unplanned or seasonal demand without cost penalties
d. *Cost*: amount spent to supply product and manage supply chains
e. *Asset management*: utilization of fixed and working assets

These attributes and their associated metrics, shown in Figure 3.9's far left column, align to *customer value add drivers*. For a supply chain design or redesign of existing supply chains, the first step in creating a new end state vision is determination of supply chain attributes based on their contribution to customer added value. The relative importance of attributes determines their associated performance metrics to drive customer value. For example, in supplying product to an OEM customer's production line, extremely high reliability performance is critical to eliminate missed production schedules and potentially losing a sale, a customer, or new sales opportunities. Failure to meet promised deliveries on time, unacceptable quality, or supplying the wrong product will result in failure to meet the customer's requirements. Poor delivery performance also affects inventory levels. Unreliable delivery causes customers to compensate by carrying extra inventory. The correct selection of supply chain attribute priorities based on customer value added results in greater EVA for both suppliers and customers. It also ensures balanced internal and customer metrics, providing effective direction for business operations. The bottom line for any company focusing on their customer's priority value-adding metrics are as follows:

a. They can become a preferred supplier as their customer's benefit from being able to service their markets efficiently and reliably.
b. Internal efficiency, cost, and growth opportunities that come with being a preferred supplier.

8. *Business processes must build a well-tuned operating model dedicated to delivering unmatched value.* Discussions so far identify Lean enterprise's significant impact on supply chain operational cost, working capital, and capital expenditures. Lean enterprise also deals with all remaining company processes and functions to achieve organizational alignment, shorter process cycle times, increased process reliability, and reduced cost through elimination of waste. This includes reducing SG&A and operational costs affected by SG&A processes and functions. Lean tools and practices are applied to transactional SG&A processes, eliminating waste and improving cycle times in a fashion similar to how they are applied in operations; the LEOMS applied to transactional business process will be covered in Chapter 8. The LEOMS is a complete operational system for any enterprise desiring to maximize its competitive potential. Toyota's visibility and success makes them a great example of an enterprise driven by value discipline and proposition.

Toyota became successful with a product leadership value discipline based on quality, reliability, and low lifetime cost value proposition. Toyota started in the United States 50 years ago competing with small affordable cars. Their initial value proposition has evolved over 50 years from offering small, highly reliable, fuel-efficient functional automobiles to meeting consumer segment needs for luxury, best value, and economical transportation, respectively. Toyota Motor Company's three brands—Lexus, Toyota, and Scion—have progressively added new customer segments supported by offering unique winning value propositions. Today they have competitive products in every major consumer automobile and pick-up truck segment in North America. During recent decades, they frequently made more annual operating income and net profit than the rest of their industry. Toyota's product leadership value discipline and propositions for each customer segment are consistently reinforced through customers' experiences with their products, dealers, service, and communications. Market leading companies like Toyota consciously design their operational model to deliver customer experience as intended by their value discipline and proposition. Toyota's conscious enterprise design, driven by their product leadership, value discipline, and key customer segment value propositions, is powered by flawless execution, resulting in sustainable market leadership. Toyota has experienced some serious product quality issues in the past few years, which they attribute to becoming so focused on global expansion that they lost control of sustaining their company culture resulting in damage to their quality and reliability reputation. After openly admitting their problems, they have moved forward and recovered their position as the most profitable car company in the world with operating margins significantly better than their U.S. competitors.

Strategic, Core, and Infrastructure Processes

Real product and service competitive advantage results from strategic business process and differentiated competency excellence managed by an organization that is always raising the bar on competitors. Strategic processes are those containing differentiating competencies and capabilities, the basis of a company's competitive advantage. For example, the study of customer applications is a part of a marketing process that understands customer use of services and products to determine requirements and trends, and is a strategic process. New requirements discovered from customer application studies can only be satisfied if production processes for building the product have capability to deliver the evolving application requirements. Therefore, manufacturing's process technologies are strategic processes. Rule 3: Maintain threshold standards on other dimensions of value. Maintaining threshold performance means you must have industry parity on all dimensions of competitive performance to avoid being at a disadvantage versus market competitors. The speed of global markets is continuing to raise the bar on minimum threshold levels, below which a company is competitively disadvantaged. As a component of Lean enterprise implementation, all operational metrics are reviewed to establish required levels of competitive performance to support a company's stated value proposition. They are categorized as requiring superior advantage or parity performance versus current levels of top market competitors. This process ensures that an organization's effort is focused on improving processes based on gaps between current performance and target performance. Producing newly designed products or services is typically a strategic process required for satisfying the customer, but in and of itself does not create differentiation unless the product is delivered on time and at an acceptable cost. Checking credit is an infrastructure process as it is a required part of running a good business and needs to be done efficiently and effectively to avoid customer dissatisfaction. Lean is applied to all enterprise business processes whether they are strategic, core, or infrastructure processes. Lean focuses on quality, reliability, and speed in strategic process with a clear definition of the future target conditions for each process. Core process may have a focus on quality, flexibility, reliability, and cost, while infrastructure processes might focus on cost, quality, and reliability. It should be understood that Lean enterprises believe their customer is king, and a maniacal focus is maintained to satisfy customers. Often companies have an aspirational value proposition that is not based in reality, as their strategic business processes and differentiating competencies do not support the stated value proposition. Customers must experience winning value propositions, and this only happens because strategic business processes and differentiating competencies are sufficient to make it a customer reality. Application of the LEOMS will improve and transform current business processes to be capable of making value propositions a reality and in the process define specific differentiating competencies, which are missing or insufficient.

9. *Management, HR, financial, and operational processes integration from suppliers to customer integrated process design.* The improvement of functional cost is a consequence of applying Lean to all enterprise end-to-end business processes (see Figure 3.10). Lean principles provide valuable guidance in designing business processes. Their application leads to designing out unnecessary cost and delays in processes. Once in operation, those processes can be continually improved to increase customer value through rigorous elimination of waste. While waste elimination is the dominant idea Lean is recognized for, it is driven by perpetual desire to improve value for customers. This means investing in features, services, and technology that customers see, as value may increase cost, but will sufficiently increase their perception of value to result in improved financial results. When applying the LEOMS to enterprise processes, it is critical to understand the type of

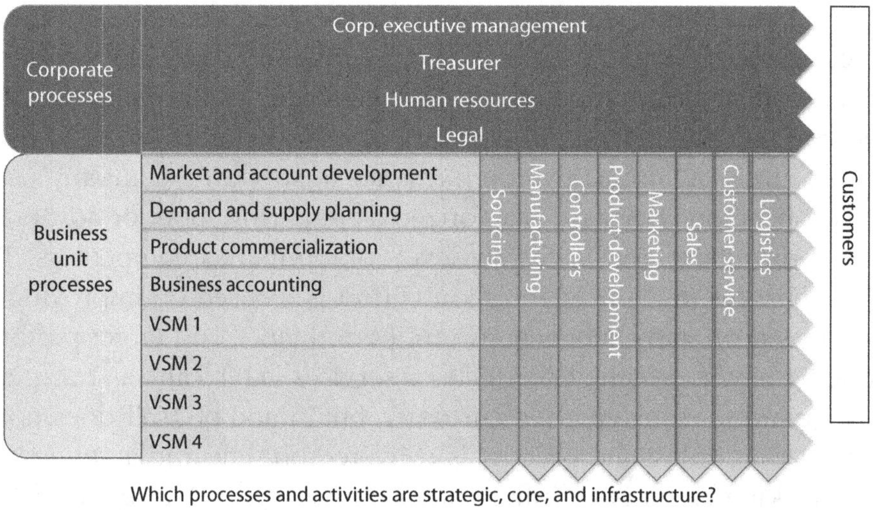

Which processes and activities are strategic, core, and infrastructure?

Figure 3.10   Business process organization matrix.

process being improved, whether it is a strategic process, a core process, or an infrastructure process.

10. *Adoption of the LEOMS.* The LEOMS is the best operational management system known today to optimize the consumption of resources in support of satisfying customers. A detailed implementation plan is presented in Chapters 4 through 8.

Summary

Just like human beings, in organizations, genes matter! Lean will have limited long-term continuous contribution to the business if it is viewed as simply an operational improvement tool set that is applied by the operational business organization. To achieve sustainable long-term continuous improvement, it must become part of the fabric of the organization, and through sustained effort be "genetically grafted" into the existing company business culture. It will take years of conscious, intentional effort to reach this point. But the rewards are great as the LEOMS can and will deliver long-term continuous operational improvement and strategic competitive advantage.

Reference

1. Michael Treacy and Fred Wiersema, *Discipline of Market Leaders*, Boston, MA: Addison-Wesley, 1995.

Chapter 4

Engaging Leadership and Preparing the Culture Change Plan

What Will I Learn?

Why and how to assess and modify, if necessary, your company's culture to be compatible with the required LEOMS continuous improvement and people leadership culture.

Assuring Organization-Wide Change Initiative Success

In the Introduction, a point was made that companies undertaking big transformational initiatives must take advice from leading researchers seriously as to why companies succeed or fail and what can be done to prevent failure. In his bestselling book, *Leading Change*,[1] John P. Kotter makes a strong coherent argument that enterprises must have a comprehensive change process to successfully and

"The challenges we now face are different. A globalized economy is creating both more hazards and more opportunities for everyone, forcing firms to make dramatic improvements not only to compete and prosper but also to merely survive. Globalization, in turn, is being driven by a broad and powerful set of forces associated with: (1) technological change, (2) international economic integration, (3) developed country domestic market maturation, (4) the collapse of worldwide communism."

John P. Kotter

permanently embed change into their organizations. He identifies four powerful forces that have created this requirement. As a part of this change process, it is critical for companies to assess and, if necessary, modify their enterprise culture to permanently sustain transformational change processes. Authors Kim S. Cameron and Robert E. Quinn, in their book *Diagnosing and Changing Organizational Culture*,[2] reported that as high as 75% of quality improvement initiatives failed to meet their goals because company cultures were incompatible with culture requirements of their chosen transformational change methodology. All companies face continuously evolving market challenges that must be overcome if they intend to grow, so the options are clear: learn to change or die! Change is not an option as globalization forces will not allow companies to stand still and maintain their market position, unless they have culturally inherent competencies and capacity to continuously improve performance and adapt to evolving market challenges. This means, based on history, that at least 80% of all companies will fail within one century unless they have strategic capability to adapt to changing markets and do the hard work of making continuous improvement part of their enterprise's daily culture. The two most common causes of failure, failing to adapt organizational culture and not applying a comprehensive change process, are addressed by the authors of the two books *Diagnosing and Changing Organizational Culture*[2] and *The Heart of Change Field Guide*,[3] providing practitioners with complementary practical validated solutions to avoid the major causes of failure. These authors provide sufficient content and tools, that a talented, highly motivated, and experienced professional could be developed into a company's own change consultant. These two books are the basis for preparing an organization to implement a successful major initiative, including transformation of a company to be a Lean enterprise. So, if your company is in a good position today, congratulations! It should still become a Lean enterprise, as it will make your position even better and continue to improve it forever. If history is any indicator, becoming a Lean enterprise greatly increases the probability of being part of the 20% of companies in existence today that will be in existence at the end of this century! A significant percent will not be around two or three decades from now. The Lean Enterprise Operational Management System (LEOMS) is not a replacement for ineffective business strategy, but it is your company's best opportunity to be in the winner's circle throughout the twenty first century because it significantly increases its sustainable competitiveness. The only question is, are you and your organization prepared to pay the price of a long journey, including changes to your company culture and a significant commitment of resources, to achieve success?

Three to Five Year Implementation Time Horizon to Achieve Sustainability: Think Long Term, Act with Urgency

It takes 3 to 5 years to develop a mature Lean enterprise that consistently improves sustainable competitive advantage and maybe as much as 10 years before it is completely inculcated into company culture deep enough that back sliding will not occur.

In the book *Corporate Culture and Performance*[4] the authors identified contributing factors in their study of high-performance organizations. The five most important behaviors in shaping excellent cultures in the successful firms surveyed were:

1. Founders clearly articulated mission and purpose.
2. Success built belief in the strategy and mission.
3. Timeless values were explicitly defined.
4. Management by-in was gained.
5. Teaching the culture was ongoing.

Conditions 2–5 are part and parcel of a kaizen culture. Management is engaged in putting these values into action. Kaizen is built on a set of timeless values and principles. The kaizen process applies scientific problem solving and creative thinking to build belief through success. If the engagement is weak, the values turn out not to be timeless, and if the results are inadequate, people in the kaizen culture do not give up but give the PDCA cycle another turn.

Behavioral and cultural changes take time to be grafted onto a current culture. This does not mean significant results will not be generated along the way; but sustainable transformation requires internalizing changes in thinking and practice, and this can only be accomplished with time. In *Creating a Kaizen Culture*,[5] Miller et al. argue that the only true test of leadership greatness is sustaining excellence at the highest level. Toyota has been perfecting the Toyota Production System for more than 60 years and has sustained quality, cost, service, and customer satisfaction leadership for decades. Top executives at Toyota understand their operational system because they worked and contributed to making the system stronger. An important thing to learn from Toyota is that companies must have their own operational system, their own version of the Toyota Production System. This system provides a common language, structure, practices, and discipline, creating an "army of improvement engineers" that make small improvements everyday focused on customer service, process improvements, product performance, and reliability resulting in reduced cost. This shared operational system enables retention of organizational knowledge and experience that is captured in standard work documents and organizational culture, preventing common problems from reoccurring as happens in many companies when one leadership generation passes the baton to the next generation. This organizationally retained knowledge prevents successive leadership generations from having to resolve problems solved by previous generations of leaders, a common experience in many organizations.

Step 1. Lean Enterprise Implementation Transformation Leadership Engagement

It is worthwhile at this point to move away from talking about the Lean Operational Management System and rename it "The YOUR COMPANY Operational Management System." This is important as a symbolic act of owning the system and beginning the process of owning its processes and practices as a core part of the company. The journey of becoming and sustaining a Lean enterprise has no endpoint, but like pretty much everything in life, starting your Lean enterprise baby with the right genes matters immensely. Because of this, the top leadership team must be fully engaged in deep analysis and examination of the enterprise's current state and future state, considering market forces that are or will affect the companies competitiveness. This process of facing reality, the truth, may require facilitation from someone outside or an exceptional high-level internal leader, who can separate their personal company connection and recognize their biases, to lead a multifunctional team in examining company sustainability given market trends and forces. This team should consist of highly talented professionals (the next generation of top leaders) from strategic planning, marketing, sales, product development, process development, finance, supply chain operations, and human resources. This is important for establishing sustainability, as the current and next-generation leaders will have ownership of the Lean enterprise transformation. Their task is to prepare a "white paper" on current and emerging market threats and opportunities and how they will likely impact company performance and longterm sustainability. This document provides the "reason to change" component of the business case. This next-generation multi-functional leadership team should continue to function as the Lean enterprise implementation leadership team and be key contributors to the vision, business case for Lean enterprise, and organizational mobilization because in the process they will become joint owners with existing top leadership of the Lean enterprise transformation.

Top Leadership's Lean Enterprise Transformation Roles and Responsibilities

CEO Must Get the Transformation Started Off on the Right Foot

Introducing a transformational change must be handled very skillfully by the CEO. Organizations hold many beliefs about the value of how things have been done and take pride in what they do every day, so CEOs need to approach communications about transformational change to avoid an unintended consequence of people in the organization taking the proposed change as personal criticism. This takes a very skilled CEO who must deal with this in a way that is respectful of the past, while challenging leaders to "do it again" by rebuilding a winning company. I saw this done at 3M by Jim McNerney, retired Chairman and CEO of Boeing.

TELLING COMPANY LEADERSHIP THAT
THEIR BABY HAS SOME ISSUES

Jim McNerny became 3M CEO on January 1, 2001 and about three months later after traveling the world to become familiar with 3M global operations held an executive meeting. He started the meeting by talking about his trip and the many things he was impressed with about 3M. After making us all feel good and showing he understood what had made 3M great, he proceeded to tell us about our reality with margins that had declined in the past few years that must be restored to support our continued investment in R&D. He told us that his experience with Six Sigma at GE proved to him that we could restore those margins. The bottom line, leaders accepted his message well as they saw the numbers and in a few months people trusted Jim McNerney.

Paul Husby

In order to get Six Sigma Program buy-in from the most talented 3M professionals, their career success was connected to taking a leadership role in the Six Sigma process. Once McNerney had all the executives feeling good, he introduced the cost each executive would have to contribute to implementing Six Sigma in his/her own business. It was not long before all executives were seeing improvement results show up in their P&Ls.

LEARNED FROM EXPERIENCE: CHANGE INITIATIVES
SUCCEED BY LINKING MANAGERS CAREER
PROGRESS TO CHANGE INITIATIVE SUCCESS

After GE's Jim McNerney joined 3M as the Chairman and CEO in January 2001, 3M became a Six Sigma company in March 2001. He personally championed Six Sigma by creating an executive leadership position reporting to him, a training and deployment infrastructure, and setting up an organization in each business that included a Six Sigma Director, MBBs (Master Black Belts) and BBs (Black Belts) Six Sigma was also promoted as an important leadership development experience and soon all the most talented people wanted to participate in Six Sigma.

Jim McNerney
Six Sigma Leadership at 3M—Paul Husby

GETTING TOP LEADERSHIP TO BUY-IN

I had the opportunity to watch a master lead the implementation of Six Sigma at 3M. A few months after Jim McNerney arrived at 3M as CEO and Board Chairman he initiated Six Sigma. As a Division Vice President I was "requested" to provide high potential personnel as my division Six Sigma Master Black Belt and a dozen Black Belts across the Abrasives Division, and do it without any replacements. It felt very painful at that moment but after we stopped whining and the Six Sigma Projects started producing results we were all very proud; after four years 3M operating margins improved 5% points.

Paul Husby
3M operating margins increased 5% points

Top Leadership Team Engagement

Top leadership must act as a team and be of "one mind" to

- Create and communicate a compelling Lean enterprise vision. A formalized communications process along with continuous informal conversations are important to engaging the entire organization and reinforcing strategy, goals, successes, and failures supporting the decision to become a Lean enterprise. Messages must be clear and compact, particularly in larger organizations to penetrate organizational layers and minimize misunderstanding as it cascades through organizational levels. Listening to every organizational level and checking their awareness, understanding, and commitment to goals and initiatives ensure leaders stay grounded in reality, as they move the organization forward to become a Lean enterprise. This is a never-ending effort as each organizational level's understanding will lead to questions and concerns about what is next and how does it affect them. Achieving complete organizational understanding requires committed, determined, persistent, and patient leadership.
- Develop and communicate the business case for change. This is the content of Chapter 1: why Lean enterprise provides information and context about sustainable companies and the recognition that 50% will last a few decades, and only 15% will last a century. This should give pause to every CEO and board of directors in every company around the world to seriously consider their company's long-term sustainability as an enterprise: a sobering realization of our global market there is no safe place unless your business model includes capabilities to continually reinvent its value to customers and address new solutions that enhance current technology and enable new technology in processes, products, and services. No company has a guarantee of an endless future, and their situations can be very different requiring different strategies and changes. For example

- The company has a long-term sustainable business model but needs increased financial resources to invest in new global markets and renew product platforms with new technology more frequently, and/or invest in more long-term technology programs that will be relevant in the future.
- The company has a mature product and technology portfolio that if not enhanced to increase profits will lead to it being in the 85% of companies that do not last a century.
- "Commodity" products dominate the company's market offerings, so without being the lowest cost producer of these products, the company's existence is in question.

Regardless of a current business's situation, becoming a Lean enterprise will make a great contribution to extending the life of current solutions and an opportunity for new technologies by creating financial resources to invest in technical staff and new technology development. In today's business environment, all companies that are serious about their long-term sustainability should be turning themselves into Lean enterprises either to create financial resources to reinvent their company or to profitably grow faster. It is not a stretch to say, become a Lean enterprise or die! Maybe not immediately but what about in the next century? Is your company doing what it needs to do today to adapt and thrive? For example

- Leading and supporting the Lean enterprise transformation as initiatives must be more than just words from the leader or the effort is doomed to failure. Inertia in an organization must be overcome, and this only happens when committed actions are aligned with management's words.
- Providing new skills training, tools, dedicated resources, and appropriate financial investments are required for organizational change initiatives to be successful. Leaders must make sure their organization has resources in place to be successful. This means challenging team assumptions and an open door for teams to come back if additional resources are required. Resources are always limited, so leaders must provide support for success recognizing speed of implementation will depend on top management support.
- Valuing leaders who are team players, team members, and providing cohesive leadership to the mission of becoming a Lean enterprise. Leaders must make opportunities to interact with all levels of their organization to communicate and check on each level's understanding and commitment to the initiative.
- Holding all levels of management accountable for progress and results in their areas of responsibility. Organizational commitment to an initiative by an organization is directly linked to leadership's demonstrated commitment. Leaders must be present, demand results, assist in overcoming barriers, keep the organization focused, praise good work, constructively criticize, and apply constructive pressure to the 10% that are not on board. There is no substitute for an occasional deep dive into the organization to understand how well the leadership chain is communicating and implementing.

NEVER BE TOO BUSY TO TAKE A REALITY CHECK

During the first year that I ran the 3M Abrasives Division, quality problems were significant enough that my boss Harold Wiens and I had regular reviews with Jim McNerney. The discussion was direct and both Harold and Jim's great experience were invaluable to me. During one meeting, Jim said he had talked with sales people during a field visit, asking if they were confident problems were being fixed. Fortunately for me, they said yes. I learned the CEO and Chairman, who had many things to improve, occasionally "stuck his finger in the cake" to get a reality check. Audit, coach and identify the next improvement!

Paul Husby
Do a deep dive occasionally

Step 2. Leadership Team Preparation for Success

Learn Proven Approaches to Managing Transformational Change

There are no success stories of radically improved results from transformational change not championed by a company's top leaders. Leaders do things they understand and believe in because that is what they have built their success on, so it is highly recommended that business leaders who are planning to take the challenge of transforming their company to a Lean enterprise read *Diagnosing and Changing Organizational Culture.*[2] Making transformational change is not a process most of us experience many times during our careers; it is not intuitive to all top leadership. For certain, making cultural changes is not a common experience, and in fact many times executives have been long-term members of the culture and may find themselves being defensive about changes that may be required. After reading this book, management/executive committee members need to select a group of next-generation leaders from all major functions, who will be responsible for developing and executing the culture change and Lean enterprise implementation plans. Together these two groups need to engage in assessing company culture and determining cultural changes needed for future sustainable success. When management committee members and the next-generation multi-functional leadership group have concluded their review of current and preferred cultures, they need to engage the entire company by cascading culture assessments throughout the organization. It is not necessary to involve 100% of the organization in the assessment process, as selected representative participants are sufficient. Reviewing and confirming culture assessments should be done with the entire organization in small groups to assure that everyone feels they hand a voice in the process. There may be versions of perspective regarding culture and what changes should be made from different functional groups and organizational levels. A view from the "bottom" of the chart, for example, may well be weighted heavily toward defining cultural changes needed to ensure all team members feel valued. This is a critical group as they are the primary engine for generating continuous improvement results. The recommended books *Creating a Kaizen Culture*[5] and *Diagnosing and Changing Organizational Culture*[2] provide clear actionable instructions, processes, and tools, some of which are illustrated in this book, to facilitate successful culture change and the LEOMS implementation process. Executive/management committee and their next-generation leadership should focus multifunctional group discussions on the "truths" discovered from the processes presented in these two books to address the right big issues that their companies need to change to develop and execute an effective implementation plan.

Leaders Must Develop a Working Understanding of the Lean Enterprise Operational System

Implementation of the LEOMS requires executive leadership to have at least a working understanding of the system. There is no evidence of significant successes without actively engaged executive leadership, and therefore, leaders must develop sufficient understanding to lead, guide, support, and coach the implementation process. How many times have we seen executives with a service problem immediately demand their organization stock more inventory, as if this is a permanent solution. Instead of implementing easy "temporary fixes," leaders must press their organization to find root causes and resolve the problems. If resolving the issue requires some time, then adding inventory for a designated time may be an appropriate temporary countermeasure, but leaders must follow up with their organization to be sure that permanent solutions are defined and implemented on a timely basis. You cannot give direction nor coach organizations if you do not have a clue about Lean terminology, processes, and thinking that explain how it works and what you should expect from the organization. By reading Chapters 1 through 4 of this book, you can gain this understanding; you may not know how to create cause-and-effect diagrams, but you will be aware of them and know when they should be used.

SIX SIGMA ROLLOUT AT 3M

The roll out of Six Sigma at 3M included all business executives being trained to the Green Belt level with the expectation they personally would do one project each year. Scorecards were maintained and reviewed by top executives to keep everyone on track and contributing to the overall annual goals. Business leaders must be fully knowledgeable and own the process.

Paul Husby

Step 3. Define Required Culture Changes

Business in the Twenty-First Century

No organization in the twenty-first century would boast about its constancy, sameness, or status quo compared to 30 years ago. "Stability is interpreted more often as stagnation than steadiness, and organizations that are not in the business of change and transition are generally viewed as recalcitrant. The frightening uncertainty that traditionally accompanied major organizational change has been superseded by the frightening uncertainty associated with staying the same" (*Diagnosing and Changing Organizational Culture*,[2] Introduction, p. 1). Peter Drucker, the father of modern management, concluded that "we are in one of those great historical periods that occur every 200 or 300 years when people don't understand the world anymore, and the past is not sufficient to explain the future" (*Secrets of a Winning Culture, Building High Performance Teams*,[6] p. 3). Lean enterprise has never been more relevant to enterprise success and sustainability than in the current business environment. Lean value streams are connected to and driven by customer expectations and requirements; therefore, when properly managed, they will continuously adapt to changing customer expectations. The potential of Lean enterprise to revolutionize a company's operational management model and system is nearly impossible to overstate, but to achieve its potential, a company's culture must be compatible with becoming a Lean enterprise to realize its powerful and sustainable benefits.

What Is Organizational Culture?

Culture is unseen; it is made of the organizational norms, rules, and behavioral expectations. It is part of an organization's genes; most of the time, when you enter a new organization, no one is going to teach you about their culture; if you are lucky, you will have a good friend or boss that will give you some tips and coach you when you "break" one of the unseen rules. Oscar Motomura, senior director of Amana-Key (a leadership development training and development firm in Sao Paulo, Brazil), published an article called Managing the Invisible,[7] a very insightful piece discussing the enormous cost of negative culture behaviors and unwritten rules when they are obstacles to organizational success and barriers to adopting new ideas. While we all are aware of these and frequently can identify them, Motomura makes them very explicit as the point of his article is the need for the unwritten rules to be confronted and changed to enable a new positive culture to be created and sustained. These behaviors will only change when a conscious persistent effort is made to change the culture. Some organizational cultural behaviors and preconceptions are not compatible with being a Lean enterprise and must be changed. Some examples from Motomura's book are as follows, the cost of an organization's:

1. Lack of harmony and interpersonal aggravations in day-to-day operations
2. Comfort based on too many resources
3. Arrogance that blocks learning
4. Comfort based on past success
5. Not using its talents
6. Cordial anarchy culture
7. Lack of coordination and collaboration
8. Inability to optimize resources

These are just a sample of cultural norms existing in organizations that have negative impacts on organizational productivity and effectiveness. It is critically important to understand the current culture and what needs to be changed to enable a successful transformation to being a Lean enterprise.

Culture as the Distinctive Differentiating Ingredient of Competitive Advantage

Successful companies have developed something special that supersedes corporate strategy, market presence, and technological advantages. Although strategy, market presence, and technology are clearly important, highly successful firms have capitalized on the power that is manifested by developing and managing a unique corporate culture. This power abides in the ability of a strong unique culture to[2]

1. Reduce collective uncertainties
2. Establish social order (makes clear what is expected)
3. Create continuity, perpetuate key values and norms across generations of members
4. Create a collective identity and commitment, binding members together
5. Elucidate a vision of the future energizing forward movement[8]
6. Powerfully affect performance and long-term effectiveness of organizations[9]
7. Have a positive impact on employee morale, commitment, productivity, physical health, and emotional well-being

What is evident from this list of strong team member attributes in positive cultures is their greater propensity for change as they have a strong commitment to and confidence in the organizational environment making them more open to new ideas and change.

NUMMI Motors Success Story

In the 1950s, General Motors (GM) had an assembly plant in Fremont, California. GM had built plants across the south and southwest as part of a "Sun Belt" strategy as most of these states were right-to-work states. The United Auto Workers Union viewed this as a union avoidance plan by GM, and they aggressively worked to organize these plants, which ultimately became some of the most hostile conflict-ridden plants in their network. The Fremont, California plant was among the most troublesome plants and manufactured the Chevrolet Nova. Fremont was an extremely large physical facility with several million square feet under one roof and was operating at an extremely low level of output in 1982. The plant environment was hostile as absenteeism averaged 20%, and 5000 grievances were filed in an average year or 1 grievance for every plant employee or 21 grievances every workday. In a typical year, three or four wildcat strikes occurred with all employees walking off the job. The cost of producing a car was 30% higher than Japanese competitors, sales were declining because of poor quality, and the plant productivity and customer satisfaction was the worst in GM's network. GM tried numerous initiatives to improve employee relations, quality, and productivity, but to no avail as nothing worked. The plant was damaging all GM brands as Fremont provided vehicles to all brands, so GM had no choice but to close the plant. After closure, GM decided to take a different approach and contacted their top new competitor, Toyota, regarding establishing a partnership to design and build a car together. GM was losing market share to Toyota and their Toyota Production System, recognized as the world's best car manufacturer. Since GM had not been able to figure out how to turn around the disastrous Fremont performance record, they shuttered the plant. Toyota jumped at this chance to establish their initial presence in the United States. GM offered the plant to Toyota with conditions, including using existing facilities, buildings could not be modified, existing equipment must be utilized, and Toyota must hire from the existing laid-off workforce starting with the most senior employees. Toyota made only one request: to be able to employ Toyota managers, and GM agreed. A year and a half later, after being shuttered, the plant was reopened as the New United Motors Manufacturing Incorporated (NUMMI). The first 2 years, Toyota produced Chevy Novas, and then it was phased out and replaced by the Toyota-designed Geo Prism and Toyota Corolla. The performance after a year and a half of NUMMI Motors operations begs these questions (see Figure 4.1): What were the cause or causes of this dramatic change in

Metric	Before GM Fremont Closure	After Start Up of NUMMI Motors	Percent Change	
Employees	5000	2599	50%	Reduction
Absenteeism	20%	2%	90%	Reduction
Unresolved grievances	2000	0	100%	Reduction
Total annual grievances	5000	2	99.96%	Reduction
Wildcat strikes annually	3–4 per year	0	100%	Reduction
Assembly costs per car	30% Above Japan	Equal to Japan	30%	Reduction
Productivity	Worst in GM	Double GM Avg.		
Quality	Worst in GM	Best in GM		
Customer satisfaction	Worst in GM	Best in GM		

Figure 4.1 NUMMI performance results.

performance? Is it just the Toyota Production System? Or is it the Toyota culture? Or possibly both? Unarguably, it is both and this will become clear as we work through the implementation of the LEOMS. A new culture requires not just the methods and techniques but also establishing an environment and management approach that, among other things, ensures all employees are valued and engaged in not just doing their job but also in developing themselves and making the system better. So, what changed at Fremont? Obviously, Toyota Production System made a big contribution, but that cannot explain the change in absenteeism or radical reduction in grievances. Restarting the plant with the "Toyota Way," changing the underlying plant culture, generated the greatest change. Factory floor associates obviously felt much more valued by Toyota than they did by GM, as their minds as well as their bodies were engaged every day, giving them greater job satisfaction, a sense of ownership, and pride in their plant (see Figure 4.1). Fremont plant's transformation is powerful evidence of an organizational culture's hidden power, an object lesson for all who embark on transformational change in their company (see Figure 4.2). How do you change the invisible? Fortunately, a methodology for changing culture has been developed over the past four decades by Kim S. Cameron and Robert E. Quinn, and they have provided a "handbook" to successfully make cultural changes. A high-level implementation process summary is presented in this chapter so readers can gain a basic understanding of the author's culture change approach, but to completely understand it requires reading their book *Diagnosing and Changing Organizational Culture.*[2] A presentation made by Toyota at one of Lean enterprises' "Lean Summits" describes very succinctly the challenge of replicating Toyota's system and its culture, "You can't copy our performance unless you can copy what is going on in our people's heads." This statement is a very clear way of saying that if you cannot replicate the attributes of our culture, you cannot replicate our performance. This is the challenge that is faced by every company aspiring to become a Lean enterprise, and the bottom line is if you cannot replicate the attributes of Toyota's culture, you will have great difficulty replicating its performance in year after year continuous improvement.

- Can Toyota Production System be copied? "You can't copy our performance unless you can copy what is going on in our people's heads."
- Toyota Production System was never primarily about the andon cord. It was about creating a "community of scientists" from regular people.
- It is hard to follow Toyota Production System, even at Toyota.
- Can operations be the source of competitive advantage (since operations can be copied)?
    - In the long term
    - If leadership doesn't continue to lead improvement

Figure 4.2 Lean is culture, processes, and methods. Notes from a Toyota Presentation at Lean enterprise "Lean Summit."

Organization Culture Types and Assessment

What Is Organizational Culture?

According to Cameron and Quinn, organizational culture has observable and unobservable elements (see Figure 4.3) that are often not thought about when trying to understand a company's performance. Some people have significant success in their careers because they are very adept at understanding the observable and unobservable culture and immediately using that knowledge to get ahead. Some people rely so much on "playing the game by the rules" that they are often thought about by their peers as being political. It is important for the "let's get something done" type of people to understand some of the important questions like, how do I get ahead around here? What are the unwritten rules? What is the best way to get along in this organization? What is true about company culture is this: one needs to understand it if they want to succeed in the company. Culture defines "how things work around here." It reflects the prevailing ideology that people carry inside their head. Culture is the social glue binding an organization together. Cultural elements include an organization's "implicit assumptions," the basis for "implied contracts and behavioral norms." An example might be whether an organization's way of getting something done is through consensus or strong individualism. Another example is "how decisions are made around here." Culture norms have a huge effect on decision-making, and it is important to recognize this if a person wants to get anything done (see Figure 4.4). We like to think that decision making is completely objective, but that is seldom the case, as organizations have built-in premises about what kinds of risks they are more likely to take. For example, development of new technology can be very risky, as thousands of scientist hours must be invested and new processes have to be invented through a certain amount of trial and error learning. For some organizations, this is normal and done routinely because new technology is their source of competitive advantage giving them a comfort level

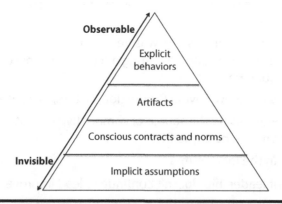

Figure 4.3 Elements of organizational culture. (From Kim S. Cameron and Robert E. Quinn, *Diagnosing and Changing Organizational Culture*, Jossey-Bass, 2011.)

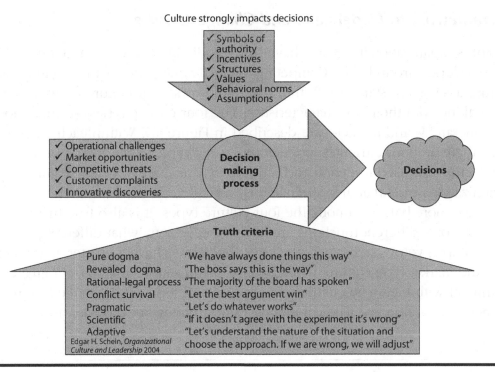

Figure 4.4 Culture impact on decisions.

about technology investment risks. Other companies without this same history and comfort in investing in new technology may struggle, debate, and ponder for a long time before taking a decision. Culture artifacts are some of the visible elements such as a company's vision, mission, guiding principles, and goals; others would include office design, location, and office furniture; another is what is celebrated as success and how it is recognized. Behavioral norms are also visible as leadership's behaviors may be one of the most reliable ways to understand culture. We often do not explicitly recognize our culture until we are placed into a different one.

THE REALITY OF CULTURE SHOCK

The first time I relocated out of the United States I experienced culture shock. The best way I can describe, it's a feeling of complete overload, like all of your senses are overloaded and you can't rationally explain your emotions and distress. Everything is different including the language so you feel incompetent as an adult. Changing company cultures may not be that stressful but one would do well to treat it like you had moved to a foreign country. You need to learn the language, rules of the road, the values, and how to successfully interact with peers, subordinates, and bosses.

Paul Husby
Culture Shock

A Framework to Understand and Change Culture

Business organizational cultures have been studied for decades, and a consensus developed around "The Competing Values Framework" as the organizational culture assessment standard. Within this framework, four culture types are identified along with their key characteristics. The four culture types—clan, adhocracy, hierarchy, and market—are described in Figure 4.5. Within each culture type, there are four attributes: orientation, leader type, value drivers, and theory of effectiveness used to define each culture in a practical way (see Figure 4.6). Organizations exist that are dominated by a single culture, and others exist that have more balance among the four culture types. It is also true that within organizations, different functions are likely to have somewhat different profiles. For example, you would expect the sales function to have a greater alignment with the market type culture, while finance organizations may have a greater alignment with hierarchy culture and product development with a culture more aligned with adhocracy. The competing values framework provides a logical,

Clan culture	Adhocracy
A very friendly place to work where people share a lot of themselves. It is like an extended family. The leaders, or heads of the organization, are considered to be mentors and, maybe even, parent figures. The organization is held together by loyalty or tradition. Commitment is high. The organization emphasizes the long-term benefit of human resource development and attaches great importance to cohesion and morale. Success is defined in terms of sensitivity to customers and concern for people. The organization places a premium on teamwork, participation, and consensus.	A dynamic, entrepreneurial, and creative place to work. People stick their necks out and take risks. The leaders are considered to be innovators and risk takers. The glue that holds the organization together is commitment to experimentation and innovation. The emphasis is on being on the leading edge. The organization's long-term emphasis is on growth and acquiring new resources. Success means gaining unique and new products or services. Being a product or service leader is important. The organization encourages individual initiative and freedom.
Hierarchy culture	**Market culture**
A very formalized and structured place to work. Procedures govern what people do. Leaders pride themselves on being good coordinators and organizers, who are efficiency-minded. Maintaining a smooth running organization is most critical. Formal rules and policies hold the organization together. The long-term concern is on stability and performance with efficient, smooth operations. Success is defined in terms of dependable delivery, smooth scheduling, and low cost. The management of employees is concerned with secure employment and predictability.	A results-orientated organization. The major concern is getting the job done. People are competitive and goal oriented. The leaders are hard drivers, producers, and competitors. They are tough and demanding. The glue that holds the organization together is an emphasis on winning. Reputation and success are common concerns. The long-term focus is on competitive actions and achievement of measurable goals and targets. Success is defined in terms of market share and penetration. Competitive pricing and market leadership are important. The organizational style is hard-driving competitiveness.

Figure 4.5 Organizational culture profiles.

Competing values of leadership culture types
Flexibility and discretion

Clan	Adhocracy
Orientation: Collaborative	**Orientation:** Creative
Leader type: Facilitator, mentor, team builder	**Leader type:** Innovator, entrepreneur, visionary
Value drivers: Commitment, communications, development	**Value drivers:** Innovative outputs, transformation, agility
Theory of effectiveness: Human development and participation produce effectiveness	**Theory of effectiveness:** Vision, innovativeness, and new resources produce effectiveness
Hierarchal	Market
Orientation: Controlling	**Orientation:** Competing
Leader type: Coordinator, monitor, organizer	**Leader type:** Hard driver, competitor producer
Value drivers: Efficiency, timeliness, consistency, and uniformity	**Value drivers:** Market share, goal achievement, profitability
Theory of effectiveness: Control and efficiency with capable processes produce effectiveness	**Theory of effectiveness:** Aggressively competing and customer focus produce effectiveness

Internal focus and integration (left axis) — *External focus and differentiation* (right axis)

Stability and control

Figure 4.6   Culture type competing values. (From Kim S. Cameron and Robert E. Quinn, *Diagnosing and Changing Organizational Culture*, Jossey-Bass, 2011.)

understandable tool set for identifying company cultures through application of an organizational culture assessment. The Organizational Culture Assessment Instrument (OCAI) can be found in Chapter 2 of *Diagnosing and Changing Organizational Culture*,[2] p. 27. This assessment is used to evaluate six culture factors: dominant characteristics, organizational leadership, management of employees, organizational glue, strategic emphasis, and criteria of success. The assessment (see Figure 4.7) evaluates the current culture and defines the preferred culture, providing definition of desired changes. Attribute statements accompanying each of the six factors are, for the most part, a way to make the invisible culture become "visible." This assessment is most useful when completed by groups followed by a discussion focused on understanding the meaning and significance of average and outlier ratings. This provides a good basis for facilitating a productive group discussion as part of confirming a group's assessment consensus. Completing the assessment is simple: within each factor are four options, A, B, C, and D attribute statements with a specific letter, such as A, which means all attribute statements are about the A (clan culture), B attribute statements are aligned with the adhocracy culture, C attribute statements with hierarchal culture, and D attribute statements with market culture. Within a set of four attribute statements, points are assigned, which must add up to 100, in a sense forcing the person filling it out to make choices. Once this is completed, scores are averaged across each culture type; the six A scores are averaged followed by B, C, and D scores using the Culture Profile Assessment Score Summary (see Figure 4.8). These average scores provide insight into the strength

Dominant characteristics		Now	Preferred
A	The organization is a very personal place. It is like an extended family. People seem to share a lot of themselves.		
B	The organization is a dynamic and entrepreneurial place. People are willing to stick their necks out and take risks.		
C	The organization is very results oriented. A major concern is with getting the job done. People are very competitive and achievement oriented.		
D	The organization is a very controlled and structured place. Formal procedures generally govern what people do.		
	Total	100	100

Organizational leadership		Now	Preferred
A	The leadership in the organization is generally considered to exemplify mentoring, facilitating, or nurturing.		
B	The leadership in the organization is generally considered to exemplify entrepreneurship, innovation, or risk taking.		
C	The leadership in the organization is generally considered to exemplify a no-nonsense, aggressive, results-oriented focus.		
D	The leadership in the organization is generally considered to exemplify coordinating, organizing, or smooth-running efficiency.		
	Total	100	100

Management of employees		Now	Preferred
A	The management style in the organization is characterized by teamwork, consensus, and participation.		
B	The management style in the organization is characterized by individual risk taking, innovation, freedom, and uniqueness.		
C	The management style in the organization is characterized by hard-driving competitiveness, high demands, and achievement.		
D	The management style in the organization is characterized by security of employment, conformity, predictability, and stability in relationships.		
	Total	100	100

Organizational glue		Now	Preferred
A	The glue that holds the organization together is loyalty and mutual trust. Commitment to this organization runs high.		
B	The glue that holds the organization together is commitment to innovation and development. There is an emphasis on being on the cutting edge.		
C	The glue that holds the organization together is the emphasis on achievement and goal accomplishment.		
D	The glue that holds the organization together is formal rules and policies. Maintaining a smoothly running organization is important.		
	Total	100	100

Strategic emphases		Now	Preferred
A	The organization emphasizes human development. High trust, openness, and participation persist.		
B	The organization emphasizes acquiring new resources and creating new challenges. Trying new things and prospecting for opportunities are valued.		
C	The organization emphasizes competitive actions and achievement. Hitting stretch targets and winning in the marketplace are dominant.		
D	The organization emphasizes permanence and stability. Efficiency, control, and smooth operations are important.		
	Total	100	100

Criteria of success		Now	Preferred
A	The organization defines success on the basis of the development of human resources, teamwork, employee commitment, and concern for people.		
B	The organization defines success on the basis of having unique or the newest products. It is a product leader and innovator.		
C	The organization defines success on the basis of winning in the marketplace and outpacing the competition. Competitive market leadership is key.		
D	The organization defines success on the basis of efficiency. Dependable delivery, smooth scheduling, and low-cost production are critical.		
	Total	100	100

Figure 4.7 Organizational culture assessment—current profile.

	Dominant characteristics	Organizational leadership	Management of employees	Organizational glue	Strategic emphasis	Criteria for success	Total	Average
Clan culture	1A	2A	3A	4A	5A	6A		
Adhocracy culture	1B	2B	3B	4B	5B	6B		
Market culture	1C	2C	3C	4C	5C	6C		
Hierarchy culture	1D	2D	3D	4D	5D	6D		

Figure 4.8   Culture Profile Assessment Score Summary.

of each culture type in the organization. All cultures will have some aspects of all four culture types, and most frequently one or two are dominant. It is important to be thoughtful regarding organizing groups of people who are brought together to participate in assessing the culture; they should be representative of the culture being assessed, not simply the "best team members." It is also useful to make assessments at various organizational levels, as there are likely to be different perspectives to bring to the process.

Organizational Culture Profile

Completing a cultural assessment generates a summary of points for each cultural type. These charts are constructed by marking the point on each of the scales that corresponds to culture type point averages from the culture assessment (see Figure 4.9). Typical profiles of organizations in various industries are shown in Figure 4.10. For example, a typical multinational manufacturer will likely have 20 points related to the clan culture, 30 points related to adhocracy, 10 points related to hierarchy,

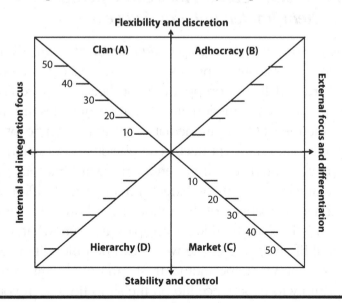

Figure 4.9   Organizational culture profile.

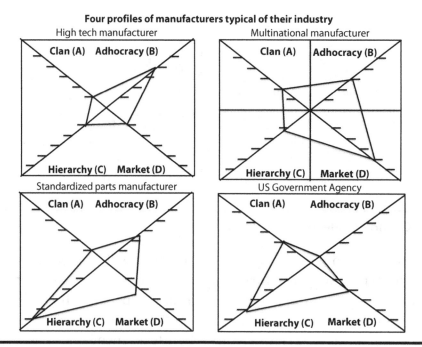

Figure 4.10 Typical industry culture profile.

and 40 points related to market culture. This makes sense as multinationals have organizations across the world supplying products and services to meet the needs of different cultures and markets across the world. Multinationals' success depends on staying close to these markets and responding to their needs. They also have to invent and/or modify products for these various cultures to meet evolving spoken and unspoken needs. Multinationals typically have limited hierarchical orientation as decisions must be made locally in support of local market and customer needs.

The Preferred Company Culture for Lean Operational Management System Implementation Success

In their book, *Diagnosing and Changing Organizational Culture*, Cameron and Quinn used the culture assessment process to evaluate visions, missions, and guiding principles of Ford and Toyota on pp. 61–63 of their book (see Figure 4.11). In the case of Ford, the vision, mission, and guiding principles are from before Alan Mulally became the CEO and immediately started a significant turnaround that has gotten Ford back on a right track. The obvious observation about the "old" Ford is the dominant hierarchy culture orientation toward internal and integration focus with some market type focus on external markets and differentiations, but no focus on innovation and creativity of the adhocracy culture. One also must consider that often visions, missions, and guiding principles are somewhat aspirational, and additionally, the real life meaning of the words can be quite different. In Toyota's case, their choice of teamwork and mutual trust can only be given meaning when one has an opportunity to experience life on the shop floor of Toyota. Prior to the arrival of Alan Mulally, Ford suffered from insufficient design innovation as

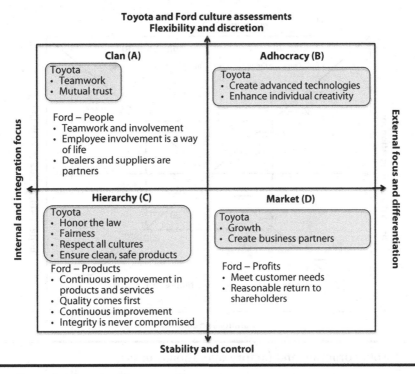

Figure 4.11 Organizational culture profile.

the cultural evaluation of Ford's vision, mission, and guiding principles infer. In contrast, when reviewing Toyota's vision, mission, and guiding principles (see Figure 4.12), there is a balance of focus on attributes of all four culture types (see Figure 4.13), an obvious hint about the best culture to enable the Lean Enterprise Operational Management System. The power of Toyota's balanced culture is confirmed with their incredibly successful introduction of the Lexus brand and its immediate acceptance as one of the top luxury brands. On the other hand, when Toyota temporarily lost its way in 2009, Akio Toyoda, president and CEO of Toyota Motor Corporation, personally took ownership of the failure and oversaw corrective action. He admitted that Toyota became so focused on growth and geographical

1. Honor the language and spirit of the law of every nation and undertake open and fair business activities to be a good corporate citizen of the world.
2. Respect the culture and customs of every nation and contribute to economic and social development through corporate activities in their respective communities.
3. Dedicate our business to providing clean and safe products and to enhancing the quality of life everywhere through all of our activities.
4. Create and develop advanced technologies and provide outstanding products and services that fulfill the needs of customers worldwide.
5. Foster a corporate culture that enhances both individual creativity and the value of teamwork, while honoring mutual trust and respect between labor and management.
6. Pursue growth through harmony with the global community via innovative management.
7. Work with business partners in research and manufacture to achieve stable, long-term growth and mutual benefits, while keeping ourselves open to new partnerships.

Established in 1992, revised in 1997. (Translation from original Japanese)

Figure 4.12 Toyota guiding principles.

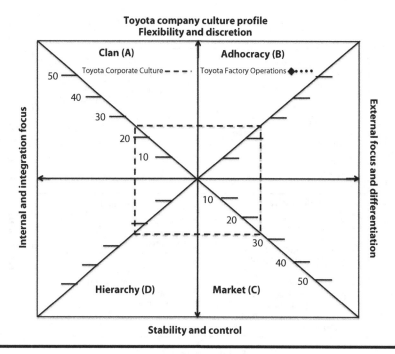

Figure 4.13   Toyota corporate and factory operations culture.

expansion that they lost their way in ensuring their culture and Toyota Production System was being properly scaled up and sustained. This was a clear indication that, when necessary, the hierarchy culture type was applied in correcting the organization's direction. He personally went public with an apology, as the failure was, in fact, a failure of leadership. Akio Toyoda demonstrated great leadership by not blaming others but facing the crisis head on and leading his organization to recovering their credibility with customers and their superior financial results compared to their two biggest U.S. competitors (see Figure 4.14). In 2016, Toyota had the vehicle industry's highest gross margins at 18%, gained 2.4% points in market share since 2011, and had the lowest warranty claims percentage and cost at 1.60%. In addition, in 2016, Toyota was the largest global vehicle manufacturer of the top six selling manufacturers in the U.S. market (see Figure 4.15). Toyota's consistency of leadership results provides validation of their balanced culture as a competitive advantage and the source of their organizational flexibility to temporarily give more

Financial and Market Comparisons			
	Toyota	**Ford**	**GM**
2016 sales	$255.528 B	$151.800 B	$166.380 B
5 year average gross profit margins	18.00%	12.26%	11.16%
Market US share 2011	12.90%	16.80%	19.60%
Market US share 2016	14.00%	15.6%	16.9%
Market share change	*+2.1%*	*−1.2%*	*−2.7%*
2014 warranty % to sales claims rate	1.60%	2.20%	2.80%
Annual cost of warranty cost	$1,040,570,024	$1,700,000,000	$2,000,000,000

Figure 4.14   Toyota's financial results versus Ford and GM.

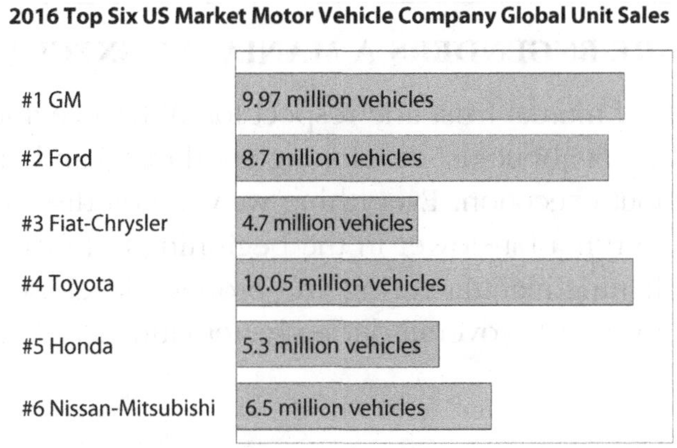

2016 Top Six US Market Motor Vehicle Company Global Unit Sales

#1 GM	9.97 million vehicles
#2 Ford	8.7 million vehicles
#3 Fiat-Chrysler	4.7 million vehicles
#4 Toyota	10.05 million vehicles
#5 Honda	5.3 million vehicles
#6 Nissan-Mitsubishi	6.5 million vehicles

Figure 4.15   Toyota has the highest unit sales globally of the top six US market leaders.

emphasis to one culture type over others when necessary and then moving back to a balanced culture when they have accomplished their objective. A logical conclusion is that the Toyota Production System is the best enterprise operational management system known to man and their culture, enables the production system as the "invisible" source that explains their sustained success.

Defining the Preferred Operations Culture to Enable the LEOMS

We typically discuss company culture as if indeed there is only one culture within a company and this is not true. While there is a dominant company culture, variations of it exist in company functions as dicussed previously in this chapter. Traditionally, we think about supply chain operations as also having a strong hierarchy culture component, as their mission is to produce consistent high-quality, low-cost products and deliver them when customers want them in the requested quantities. Toyota introduced the "TPS Paradox," an organization having a very strong clan culture component balanced by strong hierarchy and market culture components. TPS insists on very detailed standardized work methods but empowers and expects shop floor team members to continuously improve them. In other words, the structured rules for solving problems and changing standard work instructions create a "zone of autonomy," empowering shop floor team members to take ownership for continuously improving their assigned responsibility. These well-defined rules of the road create the space to grow a plant shop floor operated by "problem-solving scientists." The result is a powerfully motivated team that maintains high energy in facing great challenges. This concept was a giant step change in thinking from Fredrick Taylor's model of having shop floor team members to be only responsible for adherence to rigorously detailed work instructions that have to be completed within a defined time.

TPS institutionalized the "inverted organization chart" concept that aligns everyone in the organization in support of shop floor team members where customer value is created, a true servant leader model. Toyota's corporate culture is very

---

LEAN CULTURE ENGENDERS A MANIACAL EXECUTION FOCUS

Toyota's elements of mutual trust and respect for all people and Hoshin Kanri are all cultural. It's not about just developing good plans and giving fancy presentations, it's about execution. Everything we've been discussing is cultural. Taking the time to run a bit slower in the beginning to finish faster overall is cultural. That's aligning manufacturing and product development early on so a company can finish faster overall due to better launches. Alignment in itself is cultural.

Jerome Hamilton
It's All Culture

---

balanced as illustrated in Figure 4.16. But, when examining factory operations culture, one finds a somewhat different culture, not in conflict with the corporate culture but with a stronger clan culture component (38 points vs. 20 points), see Figure 4.16 and a lesser adhocracy culture component (12 points vs. 22 points). This culture profile was created using average cultural assessment scores from individuals with direct Toyota plant experience and a long experience implementing Lean. Additionally, while innovation and adhocracy culture are present in the Toyota Production System, it is not the organization's focus compared to their product technology research and development organization colleagues. How does this profile align with the words and artifacts of Taiichi Ohno and other TPS

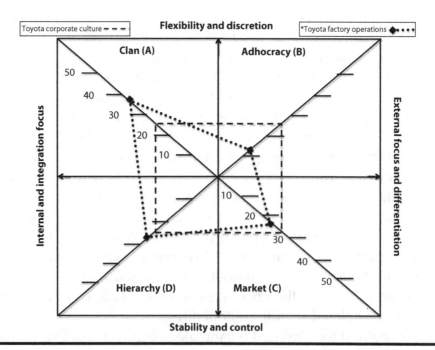

Figure 4.16 Toyota corporate and factory operations culture. (Data provided by Toyota trained professionals.)

contributors such as Masao Nemoto? The Toyota House, as shown in Figure 4.17, illustrates the level to which respect for people and teamwork are foundational to TPS success. This focus on valuing and investing in people is clearly stated by Taiichi Ohno in his book *Toyota Production System,*[10] see Figure 4.18. What stands out as the biggest potential challenge in our U.S. culture are the depth and persistence required to continuously develop team members and the patient, long-term leadership coaching required to achieve what Toyota has achieved, a workforce that is truly a sustainable strategic competitive advantage.

Even with the other elements within their culture, but without the foundational power of a "Toyota like" work force, companies will not be good enough to create the impact Toyota has shown for 50 plus years. Without building a kaizen culture, becoming a Lean enterprise will likely die or become simply a façade without reality within the organization. Getting our mind around this challenge is shown in Figure 4.19, a conceptual but relevant picture of what it would take to develop a newly hired team member into a kaizen master the Toyota way. The person develops by doing kaizen over and over again, being trained and coached but having to overcome challenges through their own initiative, creativity, curiosity, trial and error, and repetitive practice. Consider also that in addition to mastering problem solving, it requires shop floor team members to develop other skills such as the following:

- Sharing learning experiences.
- Productively participating in group problem-solving activities.
- Developing systems thinking.
- Developing greater and greater sensitivity to and recognition of smaller and smaller deviations from standards.
- Driving standards tighter and tighter to induce and expose deviations.

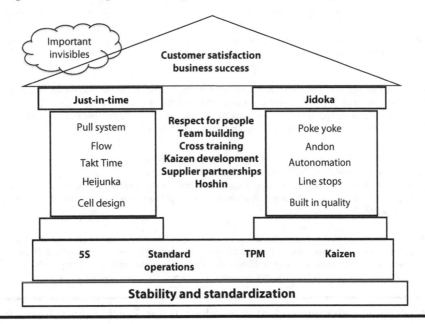

Figure 4.17   Toyota Lean house.

CREATING A CLAN CULTURE—ONE PLANT ONE TEAM ALL IN

Before I became a plant manager I was the plant superintendent and my plant manager started the practice of all salaried people spending time on the shop floor a couple of times a year. The factory operated a 4 crew 24/7 rotating shift schedule so it was easy for a big gap to occur between the salaried staff and the team members on the crews. When I became plant manager and began to lead our plant in the implementation of Lean, we instituted a new practice of running executive tours. The plant operating committee became the tour guides and first level supervisors and crew members made all the presentations on the shop floor. I admit that the first time or two I was quite anxious as I didn't know exactly how our people would do and what the reaction would be from 3M top management. In the end it was a win, win, win! Our people showed pride in what they were doing and problems they were solving, I saw ownership and commitment in our shop floor team and top management was quite pleased. In fact, the Chairman and CEO at the time had 3M's fleet of 6 Gulfstream G4's fly every executive to our plant for a tour. We hosted one group in the morning and another in the afternoon. I was pleased but the positive message this sent to our entire plant team was invaluable.

Paul Husby
Hosting the 3M Executives

Sustainable enterprise success

Create Value for Customers:

"We are now unable to sell our products unless we think ourselves into the very heart of customers, each of whom different concepts and tastes."

Stakeholder Partnerships:

"The teaching effort started with the cooperating firms nearby and spread to the Nagoya district. Toyota employees would go and help and today it is a pleasure to see all this effort bear fruit."

"Within Toyota Motor Company and its cooperating firms, people's desire to achieve the new system intensified beyond description. The system became the product of their effort."

Respect for Humanity:

"Increasing efficiency by consistently eliminating waste and the concept of respect for humanity was passed down from Toyoda Sakichi, founder of Toyota Motor Company."

"There is no magic method. Rather, a total management systems is needed that develops human ability to it fullest capacity to best enhance creativity and fruitfulness, to utilize facilities and machines well, and to eliminate all waste."

Minimize Resource Use:

"A revolution in consciousness is indispensable. Industrial society must develop the courage, or rather the common sense, to procure only what is needed when it is needed and in the amount needed. This requires what I call a revolution in consciousness, a change of attitude and viewpoint by business people. We must understand these situations in-depth before we can achieve a revolution in consciousness."

Figure 4.18 Toyota production system core values. (From Taiichi Ohno, *Toyota Production System*, Productivity Press, 1988.)

ADVICE FROM TOYOTA

"We get brilliant results from average people operating and improving brilliant processes. Our competitors get mediocre results from brilliant people working around broken processes. When they get in trouble they hire even more brilliant people. We are going to win." Since most of us are average most of the time, don't we need to work together to create brilliant processes?

Lean Enterprise Summit—Toyota Presentation

If we ponder this for a while, maybe we can begin to understand the long-term perspective we need to sustain to build a company of scientific problem-solving masters. Imagine the reassurance and satisfaction of sitting in your office knowing that today all your team members are making company products, processes, and services a little bit better. This daily continuous improvement is the only way to build a Lean culture, and it has numerous enterprise benefits. This is like a colony of ants, each carrying one grain of sand again and again and again and again until they have built their home. Of what value is one grain of sand? Almost nothing. But thousands and thousands of grains aggregated together result in constructing their ant castle. To

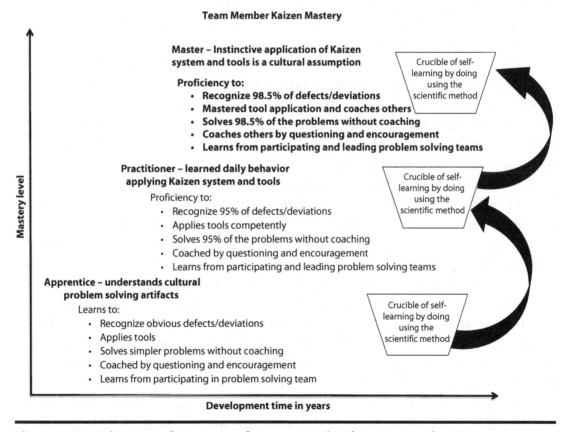

Figure 4.19   Kaizen must become a culture assumption for team members.

BENEFITS OF SUSTAINING DAILY SHOP FLOOR KAIZEN

Organizations that fail to insist on deliberate practice of dally kaizen by everyone put themselves at risk of backsliding. They will experience the gradual erosion of any benefits achieved through both daily improvements and massive improvement projects or strategic investments. There numerous benefits from sustained daily kaizen activity:

✓ Promotion of creative problem solving
✓ Shows respect for people
✓ Increases employee engagement
✓ Speeds up problem detection
✓ Reduces cost as a result of the above activities
✓ Reinforces Management credibility regarding commitment to Lean Enterprise Implementation

Adapted from Jon Miller, Mike Wroblewski, and Jaime Villafuerte,
Creating a Kaizen Culture, p. 131, McGraw Hill Education, 2014.

make this work, people must have some aptitude to become a kaizen scientist so employing "Toyota like" hiring practices for shop floor team members must assure that they have the personal and intellectual aptitudes to become "problem-solving scientists."

And so it is with the LEOMS; the methods are not complicated or hard to learn, but building a culture of thousands of kaizen engineers is an accomplishment to be admired. A case example will be developed as part of the culture discussion to make this more valuable by including two additional perspectives on what Lean enterprise culture looks like, enabling readers to complete a culture assessment of their company.

The first is a view of the culture type assessment completed by professionals with personal work experience in Toyota Production System operations. Attributes that scored 30 points or greater are presented in Figure 4.20. The consistency of these various perspectives reinforces the greatest implementation challenge, replicating Toyota's strategic competitive advantage, the Toyota workforce and the Toyota Paradox. The closest life experience to understanding the Toyota Paradox is raising children so they become successful adults. Children need structure to develop to their highest potential. They need to learn habits of successful people, such as skills to work with others, be leaders and followers, in addition to contributing ideas and action to achieve results. Having a structure allows them to explore and learn as they feel secure and loved by parents who provide structure along with the freedom to learn by doing, including making mistakes. We teach, coach, and demonstrate

3M DISKETTE PLANT GOES LEAN IN 1983

In 1983 I was the production manager and in 1984 became plant manager at the 3M diskette plant in Weatherford, Oklahoma that employed about 500 people operating 24 hours per day, 7 days per week. Our business was growing fast and becoming very competitive primarily between US and Japanese manufacturers. Prices were in free fall and our P&L operating profit was negative, we needed a drastic revolution! In 3M if your business has less than a 20% operating profit you're no longer part of the club, but we received support including investment from 3M top management. We heard about the Toyota Production System and went to a seminar that featured Taiichi Ohno, Graham Sterling and others to learn about TPS. We were just starting a complete plant process redesign project and incorporated a number of ideas we learned during the seminar. We built a 10,000 ft² clean room with a raised floor and a network of basic utilities under the floor allowing us to quickly rearrange our converting lines to support the volumes of product variations that sold as 3M branded product or OEM Brands. Product was moved through the process on carts containing a half a pallet of product. Our organization was redesigned with a shift leader and group leader focused on the main processes. JIT was implemented, we called it "Nip & Tuck" and our shop floor team members were trained to do basic preventative and minor repair maintenance tasks and had authority to stop the operation when deviations occurred. During this time Masaaki Imai* visited our plant three times and gave us advice and motivation to continue. The business did return to and sustain profitability through the products life cycle. As I look back now, I can see that we did some good things but we didn't totally comprehend the long-term continuous team member development aspect of Lean. All of the people who were there during this time still remember this period as an exciting adventure. The business was spun off in 1995 and it was the last man standing in the diskette market as the plant in Wahpeton, North Dakota was closed in 2005 as capacity and cost of integrated computer chips and Internet file transfer capability finally obsoleted the removable disk.

* Masaaki Imai tells this story in his 1986 book *Kaizen*, p. 86.
Paul Husby Leads 3M Diskette Plant to Adopt Lean

behaviors and skills almost intuitively because we were developed in the similar manner, through family culture. A second perspective is reviewing the specific Toyota culture traits listed under each of the six culture factors shown in Figure 4.21. These can also be helpful in the development of the preferred culture details.

	Dominant characteristics	Now
A	The organization is a very personal place. It is like an extended family. People seem to share a lot of themselves.	15
B	The organization is a dynamic and entrepreneurial place. People are willing to stick their necks out and take risks.	10
C	The organization is very results oriented. A major concern is with getting the job done. People are very competitive and achievement oriented.	25
D	The organization is a very controlled and structured place. Formal procedures generally govern what people do.	50
	Total	100

	Organizational leadership	Preferred
A	The leadership in the organization is generally considered to exemplify mentoring, facilitating, or nurturing.	50
B	The leadership in the organization is generally considered to exemplify entrepreneurship, innovation, or risk taking.	10
C	The leadership in the organization is generally considered to exemplify a no-nonsense, aggressive, results-oriented focus.	15
D	The leadership in the organization is generally considered to exemplify coordinating, organizing, or smooth-running efficiency.	25
	Total	100

	Management of employees	Preferred
A	The management style in the organization is characterized by teamwork, consensus, and participation.	50
B	The management style in the organization is characterized by individual risk taking, innovation, freedom and uniqueness.	10
C	The management style in the organization is characterized by hard-driving competitiveness, high demands, and achievement.	20
D	The management style in the organization is characterized by security of employment, conformity, predictability, and stability in relationships.	20
	Total	100

	Organizational glue	Now
A	The glue that holds the organization together is loyalty and mutual trust. Commitment to this organization runs high.	50
B	The glue that holds the organization together is commitment to innovation and development. There is an emphasis on being on the cutting edge.	10
C	The glue that holds the organization together is the emphasis on achievement and goal accomplishment.	20
D	The glue that holds the organization together is formal rules and policies. Maintaining a smoothly running organization is important.	20
	Total	100

	Strategic emphases	Preferred
A	The organization emphasizes human development. High trust, openness, and participation persist.	40
B	The organization emphasizes acquiring new resources and creating new challenges. Trying new things and prospecting for opportunities are valued.	10
C	The organization emphasizes competitive actions and achievement. Hitting stretch targets and winning in the marketplace are dominant.	20
D	The organization emphasizes permanence and stability. Efficiency, control, and smooth operations are important.	30
	Total	100

	Criteria of success	Preferred
A	The organization defines success on the basis of the development of human resources, teamwork, employee commitment, and concern for people.	20
B	The organization defines success on the basis of having unique or the newest products. It is a product leader and innovator.	10
C	The organization defines success on the basis of winning in the marketplace and outpacing the competition. Competitive market leadership is key.	30
D	The organization defines success on the basis of efficiency. Dependable delivery, smooth scheduling, and low-cost production are critical.	40
	Total	100

Figure 4.20 Toyota supply chain operations organizational cultural assessment. (Data provided by Toyota-trained professionals.)

Dominant characteristics

- Teamwork and partnerships
- Process and learning focused
- Systems thinking
- Secure and safe network environment
- Well defined, disciplined system
- Organization wide continuous improvement
- Long-term thinking

Organizational leadership

- Patiently coach team members to advance their professional development in applying the Lean system and not to solve problems for them as they must work through to the end in order for them to build confidence in their problem solving competence
- Success results from coaching and supporting your team
- Plan achievement
- Maintain operational discipline

Strategic emphasis

- Long-term view and thinking
- Hire people with the personal and intellectual potential for Toyota career success – Toyota People are The Competitive Advantage
- Develop the full potential of every team member
- Build the best vehicles in the world
- Develop and commercialize new technology providing customers with safer, more comfortable, efficient driving experiences
- Improve every aspect of Toyota business processes continuously forever
- Develop and sustain close partnerships with suppliers and dealers

Management of employees

People are Toyota's Greatest Strategic Advantage

- Hiring focused on the long term for Toyota and the employee
- Value people for their mind, creativity, and desire to do meaningful work
- Long-term view on developing people
- Toyota system is demanding (1) Job must be done to specific detailed instructions consistently (2) Problem solving is part of the job (3) Teamwork is an expectation (4) Understand the Toyota System
- Coach people to become competent and confident in problem solving – not "just fixing things"
- Fairness above all else based on the situation – Unusual events are treated as such and not by policy or previous precedents
- It takes years to fully develop an employee to be a fully capable team member

Organizational glue

- Hire people with the personal and intellectual potential for Toyota career success
- Develop the full potential of every team member
- Build the best vehicles in the world
- Develop and commercialize new technology providing customers with safer, more comfortable, efficient driving experiences
- Improve every aspect of Toyota business processes continuously forever
- Develop and sustain close partnerships with suppliers and dealers

Criteria for success

- Team member development and well being
- Highest quality vehicles
- Plan achievement
- Customer satisfactions (internal and external)
- Maintains and improves the system

Figure 4.21   Toyota culture.

Culture Change Example

The culture change example is provided for three reasons: (1) culture incompatibility is the single biggest reason for failure of big initiatives including acquisition integration; (2) to stimulate readers to think seriously about the absolute need for ensuring that their company culture is going to enable big initiative success; and (3) it provides an overview of key culture change processes, leadership development tools, and implementation planning.

Midwest Manufacturing Company (MMC) Culture Change Example

MMC, not a real company, will be used throughout the remainder of this book to illustrate methodologies, processes, and tools. MMC is a supplier of electromechanical components to healthcare device manufacturers and is being challenged by competitors offering lower cost products. MMC has maintained its market leadership with its superior product design and technology that customers value. Unfortunately, customers are considering qualifying MMC's competitors because of MMC's ongoing service issues, making their price premium difficult to justify. MMC's culture assessments are provided in this chapter to illustrate the culture change process. When applying it in a real situation, it is important to consider culture variations of all major company functions because the LEOMS must be adopted by all enterprise functions. The entire leadership team should be involved in providing input to assessments and building strategic action plans so they own them. Major company functions will always have culture derivations for the company culture as discussed previously. Not performing a culture assessment and developing a culture change plan will cause painful, nagging difficulties in becoming a Lean enterprise by adopting the LEOMS as an integral part of company DNA. Because company leadership is part of the culture, they must have ownership and an understanding of specifically what behaviors, rituals, and norms must be changed to increase sustainable Lean enterprise transformation success probability. This knowledge will also assist leadership in choosing appropriate approaches to applying the LEOMS in each function.

Midwest Mfg. Now and Preferred Culture Assessments

Midwest Mfg.'s corporate now culture and preferred culture assessments are shown in Figure 4.22; a review of their results creates a basis for defining and documenting culture change strategic actions (see Figures 4.23 and 4.24). Though major focus should be on culture attributes with a gap of -10 or more between now culture and preferred culture scores within each of the six culture factors, the remaining culture attributes also need to be examined to determine what should be started, stopped, or done more of or less of if they represent a potential failure risk. In reviewing Midwest Mfg.'s corporate assessment, the first attribute that stands out is hierarchy culture attribute "D" under dominant characteristics with a -20 for score. This reflects a need to implement the Lean Enterprise Operational

Dominant characteristics	Now	Preferred	Difference
A	The organization is a very personal place. It is like an extended family. People seem to share a lot of themselves.	30	15	15
B	The organization is a dynamic and entrepreneurial place. People are willing to stick their necks out and take risks.	25	20	5
C	The organization is very results oriented. A major concern is with getting the job done. People are very competitive and achievement oriented.	25	25	0
D	The organization is a very controlled and structured place. Formal procedures generally govern what people do.	20	40	−20
Total		100	100	

Organizational leadership	Now	Preferred	Preferred
A	The leadership in the organization is generally considered to exemplify mentoring, facilitating, or nurturing.	25	30	−5
B	The leadership in the organization is generally considered to exemplify entrepreneurship, innovation, or risk taking.	25	20	5
C	The leadership in the organization is generally considered to exemplify a no-nonsense, aggressive, results-oriented.	25	20	5
D	The leadership in the organization is generally considered to exemplify coordinating, organizing, or smooth-running efficiency.	25	30	−5
Total		100	100	

Management of employees	Now	Preferred	Preferred
A	The management style in the organization is characterized by teamwork, consensus, and participation.	25	30	−5
B	The management style in the organization is characterized by individual risk taking, innovation, freedom and uniqueness.	25	30	−5
C	The management style in the organization is characterized by hard-driving competitiveness, high demands, and achievement.	25	20	5
D	The management style in the organization is characterized by security of employment, conformity, predictability, and stability in relationships.	25	20	5
Total		100	100	

Organizational glue	Now	Preferred	Difference
A	The glue that holds the organization together is loyalty and mutual trust. Commitment to this organization runs high.	30	30	0
B	The glue that holds the organization together is commitment to innovation and development. There is an emphasis on being on the cutting edge.	30	25	5
C	The glue that holds the organization together is the emphasis on achievement and goal accomplishment.	30	20	10
D	The glue that holds the organization together is formal rules and policies. Maintaining a smoothly running organization is important.	10	25	−15
Total		100	100	0

Strategic emphases	Now	Preferred	Preferred
A	The organization emphasizes human development. High trust, openness, and participation persist.	20	30	−10
B	The organization emphasizes acquiring new resources and creating new challenges. Trying new things and prospecting for opportunities are valued.	30	25	5
C	The organization emphasizes competitive actions and achievement. Hitting stretch targets and winning in the marketplace are dominant.	30	20	10
D	The organization emphasizes permanence and stability. Efficiency, control, and smooth operations are important.	20	25	−5
Total		100	100	0

Criteria of success	Now	Preferred	Preferred
A	The organization defines success on the basis of the development of human resources, teamwork, employee commitment, and concern for people.	20	20	0
B	The organization defines success on the basis of having unique or the newest products. It is a product leader and innovator.	30	20	10
C	The organization defines success on the basis of winning in the marketplace and outpacing the competition. Competitive market leadership is key.	30	30	0
D	The organization defines success on the basis of efficiency. Dependable delivery, smooth scheduling, and low-cost production are critical.	20	30	−10
Total		100	100	0

Figure 4.22 MMC organizational cultural assessment.

Clan culture	Organization
What should we do more of?	
Build stronger partnerships with suppliers and channel partners	Procurement
Leaders mentoring and coaching	All leaders
Leaders listen, question constructively, and coach associates	All leaders
What should we start?	
Emphasize teamwork and participation	
Empower team members to take ownership for improving their processes	Corporate wide
Supply team members resources needed to apply kaizen to their processes	All leaders
Leaders will recognize effort, progress, and results of Kaizen	All leaders
Build real relationships built on mutual respect	All leaders
What should we stop?	Corporate wide
Cordial anarchy, just getting along, and not having open conflict	
Leaders not being engaged with their teams daily	Corporate wide
Leaders being impatient, condescending, dictatorial, and suggesting solutions to subordinates	All leaders
Hierarchy culture	All leaders
What should we do more of?	
The CEO will lead the Lean enterprise implementation	CEO
Leadership will be responsible for Lean implementation in their area of responsibility	All leaders
What should we start?	
Teams will have regular "glass wall meetings" to maintain project progress and celebrate successes	All leaders
Implementation of The Lean Enterprise Operational System across the company	All leaders
Adopt The Lean Enterprise Operational System as the company's operational system	Corporate wide
All team members will follow standardized work instructions and PDCA	Corporate wide
Teamwork will be an expectation on all team member performance appraisals	Corporate wide
Modify hiring criteria to include demonstration of Lean team member aptitudes	Corporate wide
What should we stop?	
Holding people accountable for the way the boss wants to do something or to solve a problem	All leaders
Telling team members the solution to a problem in lieu of patiently coaching them	All leaders

Figure 4.23   Culture change strategic actions to be taken.

Management System's standard work detailing job tasks and kaizen as critical elements to successfully become a Lean enterprise. This attribute is one half of Toyota's clan culture's "Toyota Paradox." The second "Toyota Paradox" element is team members' autonomy and responsibility to make their jobs easier, safer, more consistent, and more productive. Standard work's structured environment creates this "zone of freedom" for team members. Making this paradox work is enabled by clan culture's other five attributes, those with an A in front of them. It is worth noting that Midwest Mfg.'s dominant clan culture attribute characteristics are predominantly based on "family" personal relationship, while Toyota's culture reflects more "professional" respect for individuals based on Toyota's commitment to provide employees with opportunities to develop their full capabilities during their careers. This results in team members feeling valued because they have more meaningful, satisfying jobs and job security. Additionally, team members possess great employable skills beyond just actual physical labor skills learned doing their daily job, in case a layoff would occur, or they wish to relocate to another part of the country. This commitment is affirmed by clan culture strategic emphasis attribute's score of 30 related to "The organization emphasis on human development, high trust, openness, and participation persists." Toyota's commitment is to

Adhocracy culture	Organization
What should we do more of?	
Expect improvements including cost reductions from all functions in the organization	Corporate wide
What should we start?	
Apply Kaizen problem solving as the daily norm	Corporate wide
Encourage experimentation to choose the solution for a problem's root cause	Corporate wide
Teach the entire organization about creativity and out of the box thinking	Corporate wide
What should we stop?	
Applying "just do it" thinking in solving problems	
The entitlement culture where it exist in the company	
Market culture	
What should we do more of?	
Maintain focus on customers as the top business priority	Corporate wide
Provide customers with great value	Corporate wide
What should we start?	
Make decisions based on the best interest of the overall business and customers	Corporate wide
Expect teams to be responsible to achieve established goals	Corporate wide
Change will be implemented only when standard work is agreed to by the appropriate people	Corporate wide
What should we stop?	
Making changes that are not in the best interest of the overall business	Corporate wide
Make short-term expedient decisions postponing problem consequences but not solving them	Corporate wide
Leaders expecting results without regard to how they are achieved	All leaders

Figure 4.24   Culture change strategic actions to be taken.

its most important sustainable competitive advantage, its team members, as they generate financial and brand credibility that underpins company success.

Culture Change Strategic Action Agenda

Creation of Culture Change Strategic Actions to Be Taken (see Figures 4.23 and 4.24) is initiated by involving all leaders in facilitated sessions of 8–10 leaders to develop proposed actions to be taken as Culture Change Strategic Action Plan input. Group sessions start by reviewing the final corporate culture assessment to assure that all participants have a common understanding. To make this process easier, assessment culture attribute statements should be organized by culture type, so all A attribute statements (clan culture) are together followed by B, C, and D. Obviously, statements associated with a big gap between now and preferred culture ratings need the most change, but even those that have no difference should be reviewed to see if change should be considered to more completely align culture types, thus better enabling support of Lean enterprise implementation. The group should then review each attribute statement and discuss together what it means in terms of actions and behaviors. For example, see Figure 4.22 MMC Organizational Culture Assessment, clan culture (A) first attribute statement under the dominant characteristic culture factor is as follows: The organization is a very personal place; it is like an extended family, people seem to share a lot of themselves. This could mean many different things to group participants, and each of their perspectives is valued; for example

- All leaders should know the first names of all team members under their responsibility and should meet regularly with groups of people at every level of their organization.
- Leaders should have "morning coffee" with small groups of their employee on a regularly scheduled basis.
- Leaders should sponsor an organization-wide picnic and a family day to allow families to see where their family member works, what he/she does, and meet their work friends.

Once the group is satisfied, they understand what attribute statements mean; they can move on to brainstorming these three change action questions, (see Figures 4.23 and 4.24):

- What should we do more of?
- What should we start?
- What should we stop?

When groups "run out of gas" on brainstorming, they can review all identified suggestions, consolidate similar/same thoughts, and prioritize them within each culture attribute. All groups' suggested change ideas and their priority are then reviewed to become inputs for top executive teams to finalize for inclusion in the Culture Change Action Plan.

What the Change Means and Does Not Mean

Changing company culture has some similarities to moving to a different culture; one needs to learn to read the signals correctly, learn local customs and norms, learn how to get things done and how to deal with people in the best way. When we are in our own culture, we learn to read the signals whether they are spoken or not; when we are in a new culture and our language skills are limited, it is very difficult, as all we understand is the literal translation of the words we hear. So, it is important to give context to proposed culture changes to facilitate a proper interpretation of the changes. A very useful tool is to define for each change what it means and what it does not mean (see Figures 4.25 and 4.26 as an example). This provides clarity to the organization, so what is expected of

Midwest MFG. corporate culture type effectiveness and organizational theory										
Culture type: clan						Culture type: adhocracy				
Increase	X	Decrease		Stay as is		Increase	X	Decrease		Stay as is	

Means:	**Means:**
Promoting teamwork and participation	All team members on kaizen are expected to apply it consistently
Leaders are trained to listen, question constructively, and coach	Team members are expected to eliminate or control process defects
Team members are empowered to improve their process	Team members are encouraged to try (experiment with solutions)
Providing daily time to work on process improvements	Team members are expected to make improvement year over year
Encourage team member ownership for their jobs	Cost reductions must be good for the overall business
Leaders are expected to know all their team members	Encourage and celebrate risk taking
Kaizen resources will be supplied to teams	
Leaders are expected to recognize effort, progress, results	
Doesn't mean:	**Doesn't mean:**
Just getting along well and not having conflict	Kaizen does not mean we apply "just do it" thinking
Leaders being impatient, militaristic, superior, condescending	Team members can choose to improve whatever they want
Team members can do their job as they wish	Experimentation can be done without regard to safety risks
Team members can do kaizen in place of their other job duties	Because I've made more improvement than others I've done my part
Process ownership doesn't allow changes to be made arbitrarily	Cost reductions can't risk quality, reliability, service, or safety
Leaders stay in their office, isolated, and uninvolved	
Teams will get anything they request	
Leaders only care about the end result	

Figure 4.25 Example of "means and does not mean."

Culture type effectiveness and organizational theory										
Culture type: hierarchy						Culture type: market				
Increase		Decrease		Stay as is	X	Increase		Decrease		Stay as is	X

Means:	**Means:**
The CEO is leading the Lean Enterprise Implementation	Maintain focus on customers as a top priority
Adopting Lean Enterprise as our Operational Management System	Providing customers with greater value
Teamwork expectations are on all performance appraisals	Work teams will be responsible for achieving established goals
Team members regularly report on progress at glass wall meetings	All means necessary and available are to be used to satisfy customers
Team leaders will hold 5 minutes daily start of shift meetings	
Team/group leaders are responsible to train and coach their team	
New hiring criteria assuring team members have needed aptitudes	
Team members must follow standard work instructions	
Doesn't mean:	**Doesn't mean:**
Implementation leadership is delegated to business executives	Customers are a top priority except in cases causing more work
Continuing to manage operations as it is done today	Making changes that are good for the P&L but not the customers
Teamwork is a voluntary choice for team members	Work teams will be accountable for a best effort regardless of results
Team members discuss unrelated topics at glass wall meetings	"Short cuts" and questionable ethics can be used to satisfy customers
Start of shift meetings last more than 5 minutes	
Leaders train team members and expect them to apply it	
Continuing hiring people with the same criteria as in the past	
Work team members will make final decisions on what they do	

Figure 4.26 Example of "means and does not mean."

team members is well understood and does not fall prey to individual interpretation causing confusion or a tool for "resistors" to disrupt the change process.

Step 4. Leadership Training and Development

Leaders need to become familiar enough with the LEOMS methodology to be able to lead and coach their organization through implementation and sustainment. By reading this book, *along with Diagnosing and Changing Organizational Cultures, and The Heart of Change Field Guide,* and spending a week in train with a knowledgeable lean coach, executives will have sufficient initial understanding of the LEOMS and cultural change methodology to initiate their Lean enterprise transformation. They will be prepared to build a vision and business case supporting Lean enterprise implementation. Leaders' knowledge and comfort will increase as they learn to personally own and apply Hoshin Kanri, kaizen PDCA and work with their organization applying the LEOMS. Executive Lean coaches are responsible to meet regularly with all business executives to continue coaching and assisting them in their Lean enterprise transformation including regular assessments of their organization's progress.

Lean Manager's Culture Change Responsibilities

The Greatest Challenge for Lean enterprise transformation leaders will be repetitious daily coaching practice (see Figure 4.27), as it is the only way for kaizen cultures to become a reality and enable the results leaders will be expecting from the LEOMS. The reality is there are lot of consultants and Lean practitioners who can be contracted to implement the LEOMS, and it will do some good; what this will not do is establish a company full of improvement scientists that will make improvements year after year after year, forever. More important to the longterm improvement results than technical knowledge of the Lean Operational Management System is cultural knowledge, the invisible foundation and sustaining force driving longterm improvement results and the glue assuring culture changes grafted on to a company's culture will produce good fruit and a truly sustainable continuous improvement business culture. This will only happen if management

- Teach team members to focus on the work and see the value stream
- Give them deep technical and process knowledge
- Push responsibility for value stream management and improvement to the lowest practical level of line management
- Introduce metrics which encourage horizontal thinking and sharing
- Create frequent problem solving loops between managers and their bosses
- Execute the above responsibilities via policy deployment, A3 analysis, standardized work, standards management, and kaizen

Figure 4.27   Lean management coaching responsibilities—Lean Institute Executive Forum.

1. Kaizen and Kaizen again
2. Coordination between departments is an essential skill of management
3. Everyone speaks up
4. There is no reason—I do not scold
5. When the learner does not understand, try another way to teach
6. Give up your best person when doing job rotation
7. If my orders don't have a due date, ignore them
8. The rehearsal is an ideal place for training
9. Company audits that result in no top management action are useless audits
10. Top management must ask, "what can I do to help?" during audits

Figure 4.28 Masao Nemoto's management creed. Masao Nemoto was a Toyota employee and manager for over four decades. (From Nemoto Lu, *Total Quality Control for Management*, Prentice Hall Trade, April 1987.)

teams completely buy into the LEOMS and its Lean enterprise culture requirements. The management team must be totally engaged to internalize both lean culture and its operational system. Great wisdom and knowledge can be gained from reading the materials available from those who were part of Toyota or were close observers during its birthing and early development of TPS (see list of resource books in Appendix I). Masoa Nemoto is one of the TPS pioneers and offers his thoughts on coaching team members on kaizen (see Figure 4.28). Culture changes are hard work and require leaders to maintain a day-to-day commitment and discipline, leading and coaching change in their organizations management.

Cameron and Quinn provide two valuable tools for assisting leaders in making necessary behavioral changes in leadership practices. Their first is the Management Skills Profile process that is completed by leaders, their subordinates, peers, and superiors, and the second is the Culture Coach Assessment also completed by subordinates, peers, and superiors.

In the book *Corporate Culture and Performance*[4] the authors identified contributing factors in their study of high-performance organizations. The five most important behaviors in shaping excellent cultures in the successful firms surveyed were:

1. Founders clearly articulated mission and purpose.
2. Success built belief in the strategy and mission.
3. Timeless values were explicitly defined.
4. Management by-in was gained.
5. Teaching the culture was ongoing.

Conditions 2–5 are part and parcel of a Kaizen culture. Management is engaged in putting these values into action. Kaizen is built on a set of timeless values and principles. The kaizen process applies scientific problem solving and creative thinking to build belief through success. If the engagement is weak, the values turn out not to be timeless, and if the results are inadequate, people in the kaizen culture do not give up but five the PDCA cycle another turn.

Management Skills Profile Assessment

The Management Skills Profile Assessment (see Figure 4.29) is a valuable tool for both assessing leaders and also identifying and selecting leaders for future management positions. This could also be used in hiring all new team members by designing group dynamic sessions into the selection process that engage prospective team members in a series of group sessions to evaluate specific skills. Its main application is for development of management and team members with career potential as leaders. This is completed by a leader, the leader's subordinates, and superiors in order to provide information for leaders to improve skills. This management skills profile must be completed in the context of strategic action plans' defined changes related to each of the four culture types. This assures clarity of specific behaviors and actions that leaders must model to successfully lead and coach Lean Enterprise Operational System implementation and build a solid Lean culture foundation, thus assuring sustainability of the system and great results.

The instructions for completing the assessment (see Figure 4.30), scoring it, interpreting results, and applying skill improvement tips are all provided in Cameron and Quinn's book, *Diagnosing and Changing Organizational Culture* (see also Appendix III). Determining minimum skill level requirements should be

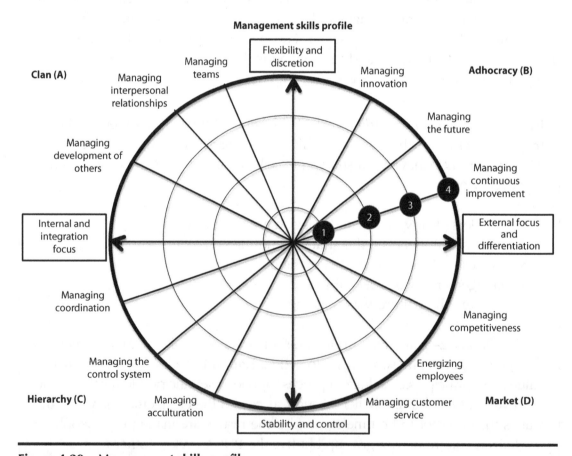

Figure 4.29   Management skills profile.

360° Lean culture coach development assessment

Sensei – Coach other coaches 156–195 Points – Continue to work on moving 3's to 5's

Coach – 117–155 Points – Continue development to moving 1's to 3's and 3's to 5's

Apprentice < 117 – Continue learning and development moving 1's to 3's

		Rarely 1	Sometimes 3	Consistently 5	Feedback
	Respect for people – leading by example				**Guidelines for suggestions:** 1. Keep suggestions positive (evaluate by thinking about how you would respond if this was given to you by others) 2. Provide specific examples (it is easiest for people to understand constructive criticism if it is a real life example) 3. Maintain confidentiality (keep your inputs to yourself and "gossiping" is not conducive to great teamwork) **Provide concrete examples and constructive suggestions for improvement**
1	Consistently treats people with dignity and respect				
2	Encourages team members to contribute their ideas in team meetings				
3	Admits mistakes and encourages team members to admit mistakes				
4	Facilitates participation of team members in decisions that affect their work				
5	Listens attentively to team members' issues				
6	Focuses team members on finding root causes and not some to blame				
7	Reaches conclusions, only after discovering the facts				
8	Gives recognition to the team's contributions to the larger organization				
9	Provides constructive performance feedback to team members				
10	Respects team members' time (i.e., starts and completes meetings on schedule)				
11	Regularly recognizes team members' contributions to the group's success				
12	Consistently rewards team members for their job performance				
	Continuous learning and development				
13	Motivates team members to continuously improve their knowledge and work skills				
14	Makes development plans for team members, insuring they receive appropriate job training				
15	Teaches all team members coaching skills and leaders how to train and coach their teams				
16	Coaches team members and groups on job performance, consistently and constructively				
17	Encourages team members to look for and expose problems so they can be worked on and solved				
18	Takes "system" perspective when looking for and making improvements				
19	Coaches and encourages a team approach to problem solving				
20	Encourages team members to share their ideas and knowledge with others				
21	Makes themselves available to assist team's from other areas in problem solving				
22	Supports team members when they need assistance				
23	Seeks out already proven solutions and promotes them with teams and team members				
24	Facilitates and encourages continuous change as good for the customer and job security				
	Lean process and methods execution				
25	Engages team members in establishing work processes and standards				
26	Regularly validates adherence to work processes and standards				
27	Responds rapidly when problems are escalated = to her/his level				
28	Ensures disciplined applications of problem solving methods when problems arise – No Just do it's!				
29	Encourages team members to stop the process when quality problems arise				
30	Requires teams to deploy visual controls in all processes when appropriate				
31	Is disciplined to take action on team members' suggestions to improve the processes				
32	Makes decisions on a timely basis				
33	Is a good example by making decisions on data and facts, not subjective information				
34	Follows up to ensure control plans are in place so solved problems don't reoccur				
35	Holds teams, leaders, and members accountable for excuting their responsibilities				
36	Focuses on external and internal customers' needs and requirements				
37	Ensures team members receive needed information on a timely basis to be able to do their job				
38	Communicates regularly the business priorities and how team members can support them				
39	Creates an open communication environment so people feel free to express themselves				
	Total				

Notes:

Area/Dept:　　　　　Team member:　　　　　Date (MM/DD/YYYY):

Figure 4.30　Coach assessment.

done by examining preferred culture profile and culture type scores to provide guidance in determining required skill levels. This is to be completed annually by peers, subordinates, and superiors to provide specific feedback on key elements of creating a Lean culture. While this may or may not be included as part of annual performance appraisals, it certainly should be seriously considered as part of advancement reviews. Remember the old saying: people "judge" us by what we do, not what we say. Skill assessments should be completed after 1 year and then every 3 years with an expectation that all leaders will apply kaizen-related management skills as a part of continuous improvement of their professional competencies, with annual progress measured by 360° Lean Culture Coach Development Assessments. Completed 360° Lean Culture Coach Development Assessment results should contribute to annual development plans for every leader, and standards for expected progress should be defined and applied using the annual assessment as input. Performance evaluations should include expectations for acceptable progress and be a significant factor in each leader's overall performance evaluation rating.

Ready to Go

The top leadership team is engaged and prepared for leading the Lean enterprise transformation over a 3 to 5 year time horizon. Next generation top leaders have created their strategic goals, business case, and culture change plans and are working together as the implementation leadership team. It is time to move on to defining leader responsibilities, building the implementation plan, training leaders and coaches, and launching the Lean enterprise transformation.

One Last Reminder

In the book *Corporate Culture and Performance*,[4] the authors identified contributing factors in their study of high-performance organizations. The five most important behaviors in shaping the excellent cultures of the most successful firms were as follows:

1. Founders clearly articulated mission and purpose.
2. Success built belief in the strategy and mission.
3. Timeless values were explicitly defined.
4. Management buy-in was gained.
5. Teaching the culture was ongoing.

Conditions 2 through 5 are part and parcel of a successful kaizen culture. Management is engaged in putting these values into action. Kaizen is built on a set of timeless values and principles. The kaizen process applies scientific problem solving and creative thinking to build belief through repeated success. If the engagement of leadership is weak, the values turn out to be just words, and if the results are inadequate, people lose faith in management. A true kaizen culture does not give up but looks to failure as a source for learning, be that in shop floor processes or management commitment and behaviors.

References

1. John P. Kotter, *Leading Change*, Brighton, MA: Harvard Press, 2012.
2. Kim S. Cameron and Robert E. Quinn, *Diagnosing and Changing Organizational Culture*, Jossey-Bass, 2011.
3. Dan S. Cohen, *The Heart of Change Field Guide*, Harvard Press, 2005.
4. John Kotter and James L. Heskett, *Corporate Culture and Performance*, New York: Free Press, 2011.
5. Jon Miller, Mike Wroblewski, and Jaime Villafuerte, *Creating a Kaizen Culture*, Burr Ridge: McGraw Hill Education, 2014.
6. Larry E. Senn and John R. Childress, *The Secrets of a Winning Culture: Building High-Performance Teams*, Long Beach, CA: Leadership Press, 2002.
7. Oscar Motomura, *Managing the Invisible*, São Paulo, Brazil: Amana-Key, 1998.
8. Harrison Trice and Janice Beyer, *The Cultures of Work Organizations*, Upper Saddle River, NJ: Prentice Hall, 1993.
9. Cameron and Ettington, *The Conceptual Foundations of Organizational Culture*, Ann Arbor, MI: University of Michigan, Appendix A, p. 152, 1988.
10. Taiichi Ohno, *Toyota Production System*, New York: Productivity Press, 1988.

Chapter 5

Lean Enterprise Transformation Preparation and Launch

What Will I Learn?

How to prepare and launch the LEOMS transformation.

Enterprise Culture (Enabler or Disabler?)

In Chapter 4, the case was made for completion of a culture assessment and implementation of necessary changes as a critical first step to successful Lean Enterprise Operational Management System (LEOMS) adoption and achievement of long-term sustainable improvement. Chapter 5 focuses on the LEOMS's implementation business case, preparation, launch and completion of wave 1. As noted in Chapter 4, 75% of large change initiatives do not achieve their goals[1], so choose to be part of the successful 15% and do not make assumptions about your company's culture; instead, take time to assess your company's culture.

TOYOTA UNMATCHED CONTINUOUS IMPROVEMENT LEADER

There is no question our factories are better than they were 20 years ago. But, after 15 to 20 years of trying to copy Toyota, we are unable to find any company outside of Toyota group companies that has been able to keep adapting and improving its quality and cost competitiveness as systematically, as effectively, and as continuously as Toyota. That is an interesting statistic and it represents the consensus among both Toyota insiders and Toyota observers.

Mike Rother
Toyota Kata, p. 4, McGraw Hill, 2010

A DIFFERENT WAY OF THINKING ABOUT WORK

"James Wiseman remembers the moment he realized that Toyota wasn't just another workplace, but a different way of thinking about work. He joined Toyota's Georgetown plant in October 1989 as manager of community relations. Today, he is VP of Corporate affairs for manufacturing in North America. In his thus far successful career (with prior factory manager jobs in several industries, Wiseman recalled they had the attitude that when you achieved something you enjoyed it. He recalls being steeped in the American business cultures of not admitting, or even discussing problems in settings like meetings."

Charles Fishman
No Satisfaction at Toyota. Fast Company, p. 82, 2006

In his book *Toyota Kata*,[2] Mike Rother adds another observation about all aspiring Lean enterprise companies wanting to have performance like Toyota; they need to adopt the LEOMS's culture requirements.

Without an appropriate culture, there is no chance to replicate Toyota's decades long trajectory of continuously improving performance. A quotation from James Wiseman captures one key characteristic of Lean's "management by process." Lean culture, specifically, "people and process management," is a key component and the most difficult to implement and fundamental to achieving a truly successful LEOMS transformation.

Lean Leadership—People and Process

Lean Leadership Is about People and Process

Our American style of management was founded on our "can do" culture and deeply instilled practices such as "management by objectives," holding people accountable for achieving their objectives. It is unarguable that this management culture produced the most powerful economy in the history of the world. This past success has often been an inhibitor to acceptance of the Lean Operational Management System's "management by process" thinking and approach and therefore has limited results in many U.S. companies who have attempted to fully embrace the system. Often, the discussion of management by process is misunderstood as not being focused on achieving established goals. Instead of thinking this way, it should be viewed as management collaborating with frontline team members applying proven problem-solving methods to achieve organizational goals. This regular engagement by management with frontline team members demonstrates that they are really "all in" with these team members in achieving operational goals. Another demonstration of collaboration is evidenced by changes in how management interacts with frontline team members when they are reviewing operations. Instead of focusing on questioning team members about performance numbers and often giving them "suggestions," the focus should be on asking questions about what defects they have found, what they have learned about their root cause(s), and what countermeasures have been implemented. The fundamental theory supporting this approach is based on creating an army of scientists who identify and permanently solve daily problems by using scientific problem-solving methods, leading to achievement of their "objectives." This engagement process demonstrates management's commitment to a relationship based on mutual respect and collaboration with frontline team members as they share responsibility for making the enterprise more competitive. Toyota's leadership approach has a very significant effect on employee loyalty, as they feel valued. Steven Spears tells about his experience with a Toyota employee talking about a seminal experience with a Toyota leader.

TOYOTA TEAM MEMBERS HAVE AN EMOTIONAL CONNECTION TO THE COMPANY

What is striking about Toyota is that when I ask people to describe a seminal experience with a leader, almost all the stories revolve around the leader doing something that helped develop the storyteller. The story is almost never about a tough call or a brilliant move the leader made; I didn't encounter the common view of managers as decision makers who tell other what needs to be done. When the Toyota people tell stories, it is not a dispassionate, academic recollection. Inevitably, at some point midway through the telling, they stop and collect themselves because the experience still has deep emotional resonance even though it happened even decades before.

Steven J. Spear
The High Velocity Edge, McGraw-Hill, p. 282, 2009[3]

The Lean Operational Management System focuses on people and process and is built upon well-defined management roles, responsibilities, processes, and practices.

Senior and Middle Management Roles and Responsibilities

Leaders are responsible for:

■ Leading by focusing on people and process to achieve performance excellence
■ Developing and communicating key strategic and operational metric targets; providing regular progress updates
■ Organization, system, and process design

- Developing team members into an army of problem-solving scientists who practice yokoten
- Creating a "run to standard and target" culture

Leading by Focusing on People and Process to Achieve Performance Excellence

Leadership roles:

- Senior leadership sets organizational direction with strategy and progress feedback loops.
- Middle management solves horizontal, cross-functional problems with joint A3 projects and goals.
- Frontline management stabilizes processes and continuously improves them through standard work, standard management, and kaizen.
- Frontline team members complete assigned duties by following standard work, diligently looking for defects and variation, using andon signals to call for assistance resolving defects when necessary.
- All levels apply plan-do-check-act and apply "management by science."

Winning by Process

Lean Enterprises Have the "Mind of Toyota"

In his book *Inside the Mind of Toyota*,[4] author Satoshi Hino starts with a 48-page chapter called Toyota Genes and DNA, tracing Toyoda family member contributions and those of other recognized contributors, from 1937 to 2000, to the evolutionary development of principles and practices that we now call the Toyota Production System. In his book *Toyota Culture and Thinking*, Mike Rother makes an insightful observation regarding the need to adopt sufficient "Toyota culture and thinking" to achieve sustained benefits from becoming a Lean enterprise through adoption of the LEOMS.

TOYOTA CULTURE AND THINKING

We have been trying to add Toyota Production System practices and principles on top of our existing management thinking and practice without adjusting that thinking and practice. Toyota's techniques will not work properly, will not generate continuous improvement and adaptation, without Toyota's underlying logic, which lies beyond our view.

Mike Rother
Toyota Kata, McGraw-Hill, p. 5, 2010

His perspective reinforces the need to carefully examine "company culture requirements" critical to sustainable success in LEOMS implementation. It is impossible to overstate the importance of taking this advice seriously.

Organization Structure and Business Process Design

The Lean Enterprise Organizational Structure

A Lean enterprise organization is one with all functions focused on operations aligned in support of the business's value streams. This principle has to be applied flexibly to reflect business scale that exists in larger companies with divisions or business units, and corporate functional structures. The objective is to get resources aligned around value streams to create focus, flexibility, speed, and responsiveness in achieving and sustaining a "mature and self-sustaining" LEOMS. The Lean enterprise business process and organizational function matrix represented by Figure 5.1 is intended as a general model; businesses must establish their organization in the best way to achieve organizational clarity and focus while not creating unnecessary business structure.

Lean Enterprise Operational Management System

Every company has an operational management system, or if they have many plants, they likely have many operational systems that evolved based on local management's experience and knowledge. Are there systems designed based on sound operating principles? Do they have an army of improvement scientists making incremental improvements every day? Do they really know if their

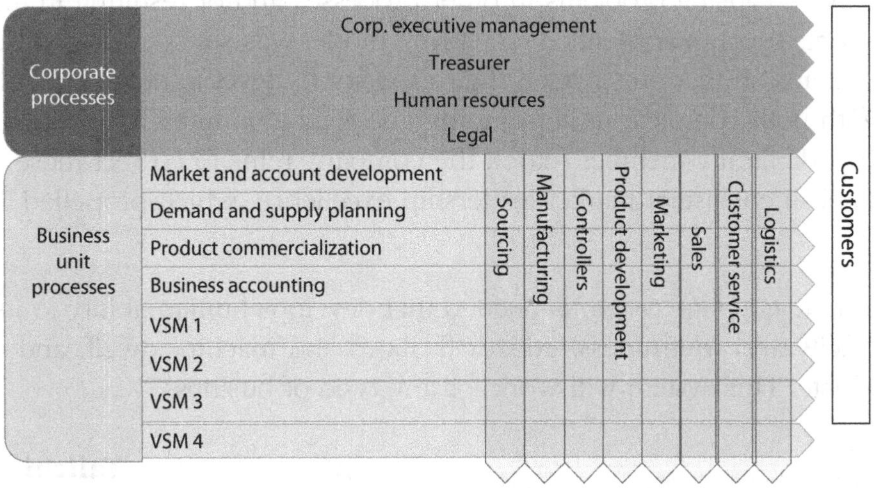

Figure 5.1 Business process and organizational function matrix.

EVERY COMPANY HAS AN OPERATIONAL SYSTEM OR SYSTEMS

"Every Company has an operational system or systems but are they able to know their system(s) are operating at a high level and improving performance everyday?" Is their system being monitored, controlled and improved every hour of every day?

Paul Husby

systems are performing at a high level every day? Are they able to detect and correct faults in their system when they occur? Are their costs improving every day? Like most geniuses, Taiichi Ohno was able to distill a complex system down into understandable principles for system design, operation, and continuous improvement, enabled by practical proven practices aligned with the LEOMS's core values.

One can spend endless hours with his book *Toyota Production System*,[5] rereading it and continuing to find new insights about the Toyota Production System. His models are made using simple figures, such as the Toyota house to communicate key fundamentals about Lean. There is also enormous implicit knowledge that comes with the Toyota house. Supply chains are complex systems that behave like organic systems, as changes in one area of a supply chain will reverberate throughout the entire system. Toyota has a long history of training new leaders by having a sensei spend dedicated time teaching and coaching them one on one. This is valuable to gain a holistic understanding of the LEOMS and how its principles and practices are connected to the whole system. One can conclude that Taiichi Ohno did this with clear intentions to ensure that new leaders understood the LEOMS and not just its observable practices. New learners need to end up with this holistic understanding so they can effectively implement the LEOMS. They need to focus on truly improving the whole and not just pieces to avoid causing problems in other processes and/or resulting in limited improvement. Toyota's apprentice mentoring model was stressed during decades of rapid expansion that outstripped their capacity to develop new generations of leaders with both complete understanding and total commitment to the Toyota way. A Toyoda family member retook the company reins to correct these errors and reestablish consistent market leadership excellence, which propelled Toyota

"A total management system is needed that develops human ability to its fullest capacity and fruitfulness, utilizes facilities and machines well, and eliminates waste. This system will work for any type of business."

Taiichi Ohno
Lean Enterprise Operational Management System TPS

to become the number one vehicle manufacturer in the world. The LEOMS is to be applied as a system, not a set of useful tools. Taiichi Ohno validates this in his book, *Toyota Production System*.

Figure 5.2 presents a more comprehensive model of the LEOMS. This system is sustained through continuous leadership training, enabling leaders to teach and coach their organization to achieve excellence in daily system execution. The system representation defines system purposes in the center circle with nine components surrounding the circle illustrating how the purpose is achieved. The nine components were previously presented in Chapter 1, but for convenience, they are repeated in this chapter.

1. *The LEOMS core values* have been in place for decades at Toyota, and unlike many companies, their stated values are imbedded in their operational model and evident in leaders' and team members' daily activities. The great challenge for companies who seek to adopt the LEOMS is developing a culture open to embracing and integrating these values sufficiently into daily behaviors to make them become a daily reality.
2. *The Enterprise Operational Management System's guiding principles* are 12 core system fundamentals starting with a focus on delivering *customer value* and shareowner returns. This is accomplished by achieving "True North," the vital few short- and long-term targets that are achieved through a contract or bond

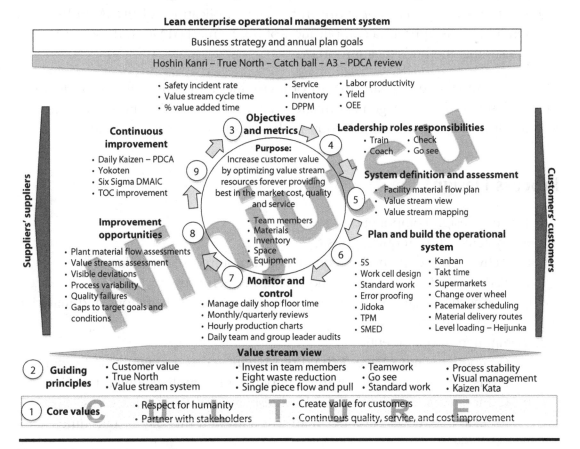

Figure 5.2   LEOMS overview.

among all organizational members to achieve them. The valuestream system provides a logical view of processes making the same or similar products spanning from suppliers through to end customers, and it is analyzed to identify and eliminate waste thereby continuously improve the system.

3. *Hoshin Kanri*, the process that establishes True North, the vital few metric targets, guides business organization's focus and effort. Frontline operational metrics are subordinate to these vital few metrics and are built to measure, control, and improve frontline operations to achieve True North.

4. *Leadership roles and responsibilities* related to system execution and improvement include regular engagement with frontline team members to train, go see, check, and coach them to control and continuously improve their processes and develop their skills.

5. *System definition and assessment.* Value stream mapping is used to define and regularly assess value streams to sustain continuous improvement.

6. *Plan and build an operational system* by applying LEOMS's practices and tools.

7. *Monitor and control* processes to detect deviations from standards and system targets, target conditions, and goals.

8. *Improvement opportunities* result from continual value stream assessments along with monitoring and control activities focused on detecting process defects and variability, in addition to performance gaps to target conditions and performance.

9. *Continuous improvement* is driven by daily applications of PDCA and yokoten as well as appropriate application of Six Sigma DMAIC and theory of constraints methods as appropriate.

Ninjutsu in Japanese means the art of invisibility and is used in the context of TPS to describe the "management magic" that appears as skills are demonstrated as a result of training the organization in the Toyota Production System.

Process Design*

1. Process design components:
   a. System output requirements.
   b. Activity or the method of completing a required set of tasks.
   c. Pathway defines the order and owner of each activity.
   d. Connection (handoff between work stations or work stations and material supply locations).
2. Component requirements:
   a. Components must have defined prespecifications and target performance levels, for example, output in parts per hour compliance with takt time.
   b. Components must have embedded tests and/or countermeasures, for example, good parts per hour this shift and this day or takt time

---

* Process design will be dealt with in more detail in **Chapter 6**.

countdown. Failure detection or current performance measurement needs to be "real-time" failure detection.

   c. Component ideal case/state—the attainable case/state on path towards "True North." It could be an intermediate case/state analogous to intermediate future states in an iterative process toward the idealized end state.

3. Key points to remember when designing processes:
   a. Outcomes must align with "True North."
   b. Design work to see abnormalities, find root causes, and take corrective action.
   c. Three common process design failures:
      i. Lack of specification
      ii. Lack of embedded test or countermeasure
      iii. Lack of specification to current standard and/or to the ideal state

4. Define and execute validation of activities per specification of work being performed by every team member; this is frequently done by group or shift leader by confirming the operation's training matrix board or chart defining who has been trained for each task and who is currently assigned the responsibility for executing the task.

5. Process design connections are signals between steps in a process and at points of integration of one process with another. Requesting material with kanban cards is an example of a connection with another process. Other methods could involve the use of a pager, an andon light, or a horn, but their purpose is the same, signaling the need of material, group leader support, or team leader support to maintain product flow in compliance with hourly output targets. Whatever method of making requests is used, there also needs to be a test to ensure that they are met satisfactorily. Signals may also be used in combination to indicate need urgency, allowing sources to prioritize, allowing them to attend to the most imminent threat to disrupting product flow.

Problem Solving

- Two types:
  - Abnormality detected when a prespecification is compared to embedded test/result
    - Stability, C_p
  - Improvement of prespecification toward ideal state
    - Improving capability, C_{PK}
- Problem solving keys:
  - Upon visual abnormality, "swarm the problem"
  - Perform rapid experimentation
  - Favor actual observations of problems
  - Stop/fix problems within takt time
  - Take determined action to improve toward the ideal case

PDCA—Plan–Do–Check–Act

A value stream manager is assigned VSM improvement responsibility for building and improving the value stream system. In addition to system-driven improvement, continuous front line operational improvement is made daily, eliminating waste caused by deviations from standards and implementing suggestions for improvement. Deviations include not achieving daily production goals, operations not consistently achieving cycle time, inventory levels not in compliance with standards, yields below standard, along with many other possible sources of deviations. Team leaders and front line associates utilize PDCA and five whys (Figure 5.3) to

- Identify the real problem
- Determine its root causes
- Define a countermeasure to eliminate or control the root causes
- Implement a countermeasure
- Check countermeasure effectiveness
- Repeat the problem-solving cycle if original problem still exists

The top-down value stream system improvement and work station bottoms-up waste elimination are two connected improvement cycles that continue driving value stream performance toward perfection. Top-down systems improvement driven through Hoshin planning assures direct alignment with key customer and company direction, priorities, and goals. The bottoms-up waste elimination cycle ensures that improvements are made every day by frontline associates who are empowered to solve their daily work problems and drive their work activities toward perfection. The result is a powerful continuous improvement process

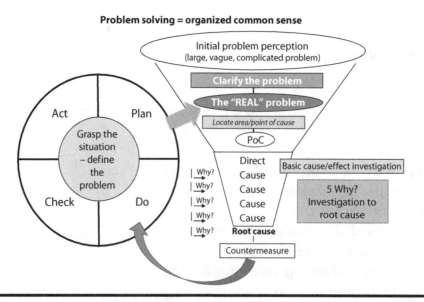

Figure 5.3 LEOMS problem-solving process.

focused on improving a company's operational and strategic competitive position every hour of everyday.

Frontline Management and Team Members

Team Leader Role

Frontline team leaders are at the point of the spear, and without high performing qualified people in these positions, building and operating the LEOMS will not succeed. Frontline leaders are responsible for a great percentage of team members in any supply chain organization as they lead team members who make and supply an organization's products. The leader has three major responsibilities: teach, support, and lead his/her team (see Figure 5.4). The leaders' teaching role is critical to team members' LEOMS culture development and maturity as they gain a greater understanding and practice daily habits of continuous improvement; this is the second most difficult and beneficial contribution of the LEOMS (second only to the challenge of adopting the leadership culture). Implementing the LEOMS takes time and knowhow, but develops a deep culture through consistent day after day, week after week, month after month, and year after year of concentrated development involving every team member. This is the only route to sustainable success and requires daily execution of team leaders' "leaders as teachers" responsibilities to build an organization of scientists, running improvement experiments daily. Only through experiments that are run daily by applying the scientific method will PDCA become second nature. Each of these experiments may only make a "meter"

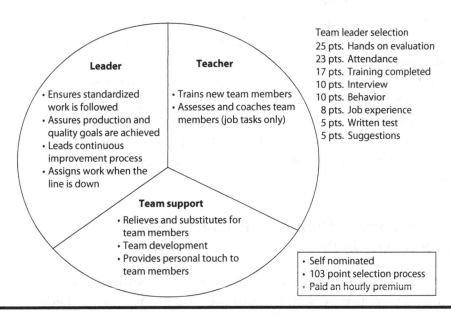

Figure 5.4 Team leader roles and responsibilities.

of progress, but when practiced by a thousand "scientists," this moves an organization a kilometer ahead each day. In 3 years, an organization has moved a thousand kilometers, and daily applications of PDCA have become the norm. This can be a great challenge in our American culture as we are frequently very impatient and want something done right away. The LEOMS implementation requires leaders to be patient servant leaders. Team members are expected to diligently focus on defect detection and apply scientific problem solving as their daily job responsibility; they are held accountable and coached by their group leader, team leader, and area manager. This leadership chain of responsibility and accountability is required to be sure that these responsibilities are executed correctly and consistently, and ensure that team members are developing their knowledge and skills. Lean Enterprise Institute identified six people engagement elements[1] for leaders to apply:

1. Teach them to see the process.
2. Give them problem-solving PDCA skills.
3. Push responsibility to the level doing the work.
4. Introduce end-to-end metrics.
5. Create frequent problem-solving loops.
6. Make abnormalities immediately visible.

Team Member Role and Selection

The LEOMS developed by Toyota is unique in its scope and depth, requiring frontline team members to perform responsibilities normally reserved for supervision, quality, and technical professionals; in addition, they have an ownership role and a voice in their workplace along with being responsible for making it better every day (see Figure 5.5).

Because of frontline team member job requirements, the employment selection process is very thorough to be sure that new team members have the aptitude to successfully execute their full position responsibilities and the potential to advance to leadership positions (see Figure 5.6). TMMK's selection process required them to start with 100,000 applicants to find 3000 candidates that met their hiring criteria.[6] *Toyota Culture*, written by authors Liker and Hoseus, is an excellent resource regarding core topics to build and sustain LEOMS's culture, and it is covered with sufficient details for experienced professionals to be able to understand and apply them.

Getting a Financial Return by Creating a Culture through Investing in People

It is fair to say that the overwhelming evidence, if reviewed by any rational, informed, and objective person, leads to the conclusion that the LEOMS is the best available operational management system for any business. Investing in

1. Set their own goals
2. Achieve and maintain perfect attendance
3. Follow standard work:
   1. Do the job right
   2. Stay within takt time
   3. Work in a safe manner
   4. Follow the same sequence
4. Follow instructions
5. Know all the aspects of the job:
   1. Tally sheets
   2. Terminology
   3. Repair ticket
   4. Specifications/C.V.I.S.
   5. Options
   6. Equipment functions
6. Be self motivated
7. Be responsible for your work area
8. Safety
9. Maintain a positive attitude
10. Be able to accept and support change
11. Maintain high standards
12. Communicate with: (be tactful)
    1. Process team members
    2. Team and group leader
    3. Manufacturing management
13. Be involved to:
    1. Eliminate muda
    2. Continuous improvement – Kaizen
    3. Submit suggestions
14. Participate in problem solving
15. Be on time and ready to go to work
    1. At your station
    2. Have all job material
    3. Have gloves, tools, pens, stamps, etc.
16. Follow the code of conduct
17. Attend quality meetings
18. Attend all team meetings
19. Train and assist in training others
20. Willingly rotate based on quality and training status
21. Be a team player
22. Verify your time card before signing it
23. Follow vacation procedure
24. Be able to work with minimal supervision
25. Take pride in your work and in the product
26. Be willing to work overtime:
    1. To support repair
    2. To support production
    3. To support training
    4. To support quality problems
    5. To attend team group meetings

Figure 5.5   Frontline team member role.

Job selection dimensions[6]

Team orientation:
- Uses appropriate interpersonal styles and methods in helping a team reach its goals
- Maintains group cohesiveness and cooperation and facilitates group process
- Provides procedural suggestions when appropriate
- Has awareness of needs and potential of others

Initiative:
- Proactive, doesn't accept the status quo and takes action maintaining active attempts to reach personal or team goals, often going beyond what is normally expected to achieve goals
- Seeks information needed to do a job and takes action without being told
- Feels ownership for the job and accepts responsibility for work outcomes and team effectiveness
- Supports team members and works collaboratively with team members to correct issues affecting the job

Oral communications:
- Effectively expresses ideas and information in individual and group situations
- Uses both verbal and non-verbal communications skills appropriately
- Demonstrates active listening skills

Problem identification:
- Identifies individual or group issues and problems
- Collects relevant information through fact finding and data collection
- Identifies plausible cause and effect relationships and collects data from multiple sources

Problem solution:
- Develops alone and with others, alternative courses of action and makes decisions
- Demonstrates use of information based on logical assumptions
- Proposed solutions are consistent with organizational values, culture and resources required

Practical learning:
- Able to follow directions and learn quickly
- Easily assimilates and applies new job information regardless of complexity
- Demonstrates reasonable and prudent attention to instructions and directions conforming to company policy

Work tempo: Consistently perform repetitive tasks at a specified tempo in a defined sequence and not exceed the normal level of waste in time, materials and supplies

Adaptability: Maintain effectiveness in varying environments, tasks, responsibilities, and team environments

Mechanical ability: Able to accomplish basic mechanical tasks

Figure 5.6   Toyota's job selection dimensions.

**LEAN CULTURE INVEST IN PEOPLE TO GET
A RETURN FROM THEIR APPLIED KNOWLEDGE**

One of the biggest elements of cultural change is not just paying hourly workers for their hands but their minds as well. Getting rid of the eighth type of waste "Lack of return on knowledge". Then training the entire workforce on Lean principles and practical problem solving. Which is what NUMMI did!

Jerome Hamilton
Investing in People to Get a Return

---

building the army of problem-solving scientists is the only proven route to building sustainable continuous improvement.

Earlier in this chapter, there is a "callout" quote from Mike Rother; to paraphrase his quote: doing workshops is typically done especially during the LEOMS implementation and making system level improvements, but daily problem-solving routines are the primary means for continuous improvement. This requires continuous training and development, a long-term investment with a great return.

Lean Enterprise Operational Planning and Execution Management

Hoshin Kanri—The LEOMS's PDCA Business Planning Process

While there are some recognized companies who seem to have mastered turning strategic intentions into effective action and results, they are the exception, not the rule. Toyota, Honda, and Danaher have shown consistency over time demonstrating this capability. That leaves most companies with the task of trying to answer some "big questions" related to strategic plan implementation effectiveness:

- How do we deploy strategic plan targets level by level?
- How do we create alignment around strategic cross-functional goals during operational plan deployment?
- How do we encourage fact-based analysis and discussion around critical strategic targets to improve (TTI)?
- Do we discuss the invisible structures, practices, or rules in our organization that may hinder effective strategic plan target achievement?
- Do we have a consistent track record of achieving annual strategic and operational TTI?
- Do we have a consistent track record of achieving the 3–5 year strategic TTI?

"I can leave the Company's strategic plan on a plane and it wouldn't make any difference: Our success has nothing to do with planning. It has to do with execution."

Richard Kovacevich
Wells Fargo CEO

Toyota's Hoshin Kanri is a proven process that is being adopted by companies in pursuit of becoming a Lean enterprise. Understanding why and how this process can effectively improve a company's operational plan deployment requires one to start by understanding ineffective plan deployment root causes.

Effects of Misalignment

Once business strategies are set, Lean enterprise organization structure, objectives, programs, and resources must be completely aligned to achieve strategic goals to minimize wasted resources. Hoshin Kanri provides the rigorous process for hardwiring application of organizational resources to achieve business strategic goals and operational targets does not exist in most companies and is the main root cause of failure to consistently and effectively achieve strategic goals and operational targets. There are three typical conditions that exist in companies related to alignment (see Figure 5.7):

1. No organization vision
2. Organization vision without departmental and individual alignment
3. Organization vision with departmental and individual alignment

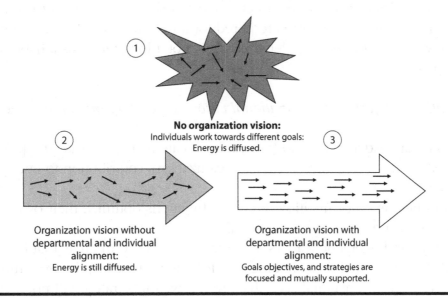

No organization vision:
Individuals work towards different goals: Energy is diffused.

Organization vision without departmental and individual alignment:
Energy is still diffused.

Organization vision with departmental and individual alignment:
Goals objectives, and strategies are focused and mutually supported.

Figure 5.7 Organizational alignment.

"The key point of this reflection is: What are we really managing in organizations and in a country? The management of what? It seems that the primary focus of management is still strongly linked to the visible, to the concrete, to the quantifiable. Management's big challenge seems to be discovering how to act on both the invisible and visible. The invisible in the broad sense, that goes way beyond what is traditionally considered intangible. Doesn't the essence of organizations and a country lie in exactly those invisible aspects of their being? The big challenge for leaders is to bring the invisible to the table where decisions are made and into people's daily lives."

Oscar Motomura
Senior Director Amana-Key, "Managing the Invisible"

Each of these conditions have organizational effectiveness and productivity consequence, but embracing the third condition is the only way an organization will achieve its strategic goals and operational targets.

Condition 1. Effects of no organizational vision:

1. Without a compelling vision, organizations have no clarity about the company's source of competitive advantage and value proposition.
2. Team members without an understanding of why and how their business wins in the marketplace have no way to connect what they do with winning. They are employed but not energized and focused on improving competitiveness daily.
3. Without differentiation based on a value proposition and the differentiating competencies to make it real to customers, a company's long-term survival is at risk. Competitive cost is required for any value proposition to sustainably win, but competing solely on cost is not a long-term sustainable value proposition.

Condition 2. Effects of organizational vision without organizational alignment:

1. Functional leaders with good intentions create metrics, goals, and programs based on how they interpret their organization's contribution to enterprise success.
2. Since there is no compelling direction, functions optimize their own area but not total enterprise customer value.
3. Creative and energetic leaders may create excellence in areas that have no real strategic value, thus creating cost for which there will be no return.
4. Strong vertical functional silos that are not well integrated will not consistently deliver value to customers.

GETTING LEADERS ON BOARD WITH HOSHIN KANRI

As I was leading the Hoshin Kanri roll out for a major corporation, I received a meeting notice from a talented Divisional Vice President to discuss the topic. When I arrived at her offices she asked me, "How do I define value to my team? I have a million things going on." I asked her to list for me the four or five key business objectives for her division, and then I asked her was her team fullyaware, and had they begun developing improvement priorities to realize the objectives. She smiled at me and said let me call a few members of my operating committee in to discuss this issue. She called in her Manufacturing Director, her Master Black belt, her Controller, and one of her Business Managers. When they arrived, she stated, "Jerome has asked me what our top four or five business objectives are, what do you guys think?" I can tell you that it was a great experience for me watching the team go back and forth in discussion for a half hour, in which after some disagreement, they were able to agree on two things that they thought everyone should have on their radar screen. She then asked her team to leave the office so she could finish her meeting with me. When she turned back towards me, I stated, "That's the value of the process. The robust dialogue that takes place that gets people aligned around accomplishing the vital few objectives." After this experience, this leader became a major supporter of the Hoshin Kanri process. Too many times we get caught up in the tools without realizing what we are really trying to accomplish. At the end of the day it's about execution on the key objectives that will increase profit and shareholder value. Even when we use the terminology "Lean Manufacturing", we should have the understanding that it is a business system.

Jerome Hamilton

Condition 3. Effects of organizational vision with departmental and individual alignment:

1. The business vision is well understood by all organizational team members. Team members understand the company's value proposition; how and why their business wins in the marketplace is clear to them.
2. Organizational leaders have metrics and goals aligned to continually improve their company's value proposition. They have clarity of how their function contributes to enterprise success allowing them to create competencies and capabilities increasing organization strength.
3. Organizations feel connected and are contributing their part to winning as team members feel valued by the organization. Being a valued contributing part of success engenders team member commitment and energy, resulting in high motivation and productivity.

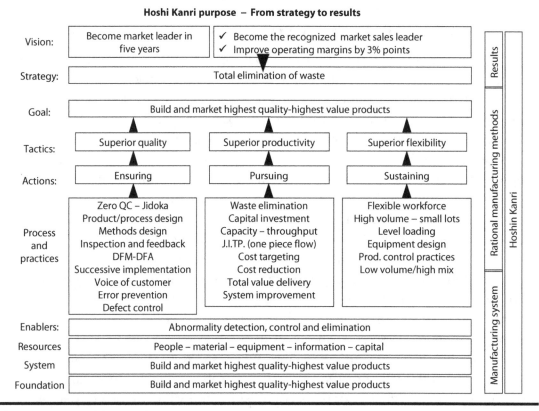

Figure 5.8 Hoshin Kanri purpose.

The majority of companies are in either condition 1 or condition 2, with a very small percentage fully meeting condition 3. One common issue is having excessive or insufficient business metrics and objectives, which are not well linked to or directly driven by customer values. This is the most obvious opportunity in most organizations to significantly improve effectiveness and productivity by applying Hoshin Kanri (see Figure 5.8).

The Solution—Hoshin Kanri Purpose

Hoshin Kanri basics:

- Originated by Toyota (Toyota Production System)
- Hoshin = a course, policy, plan, aim
- Kanri = administration, management, control, charge of, care of
- Hoshin-Kanri = policy deployment with method
- Hoshin Kanri (definition): The process used to identify critical business needs, apply PDCA countermeasures to reach defined goals and targets, develop employee capability, and align all company resources to achieve established goals and targets
- Answers the critical question "Is the ship in a storm going in the right direction?"
- Makes it possible to get away from the status quo and achieve breakthrough performance improvement

Why Hoshin Kanri?

Hoshin Kanri or policy deployment with method is an effective strategic and operational alignment of organizational metrics, programs, and resources because it will

- Identify the vital few strategic goals/targets
- Level load resources
- Engage all organizational levels through the catch ball communication and alignment process
- Align short- and long-term strategic objectives
- Establish a rhythm of periodic reviews
- Build rigor and discipline via PDCA, creating a closed-loop execution management system

Thus, Hoshin Kanri addresses root causes of poor organizational alignment by linking the vision to goals and programs across all organizational levels. This is achieved by applying its five elements:

1. True North—identify the vital few targets
2. X-matrix—integrated plan on a single page
3. Catch ball
4. A3 thinking process
5. Periodic review process with PDCA discipline

True North

The Hoshin Kanri process is initiated by an organization identifying its True North. True North is a rigorous process for defining the vital few highest priority business needs that must be achieved for success. It includes

- Business needs that must be achieved
- Gaining organizational commitment through an implied contract or bond, not merely a wish
- Macro: long-range key goals (3–5 years)
- Annual: big picture checks
- Having a micro focus on hourly, daily, and weekly execution

Defining True North is a collaborative process involving the entire top leadership team. Concurrently, top leadership is engaging their leadership teams to gain their perspective and knowledge of business needs such as critical capabilities, gaps, and barriers to success. Multilevel involvement not only ensures a better plan; it also begins to gain organizational commitment. The final outcomes are defined by the vital few metrics at all levels and full organizational commitment, a mutual bond or contract to achieve the vital few targets. The vital few include longer horizon, 3- to 5-year targets to meet future market demands by building new or stronger strategic capabilities. Finally, a big picture check is used to test the total plan, ensuring that it meets all necessary and sufficient conditions to achieve annual plan TTI and long-term strategic breakthrough program objectives.

Lean Enterprise Planning and Execution Management

The LEOMS's planning and execution process is constructed on the PDCA model providing a continuous cycle of taking corrective action on plan elements that are not achieving their TTI, as shown in Figure 5.9, to provide the greatest opportunity for organizations to achieve their strategic and operational plan goals. The process is initiated with the organization's X-matrix followed by deployment using A3 documents that are then summarized on the high-level tracking process called Bowler that is used to monitor progress at frequency intervals defined to provide early detection of deviations that require corrective actions to get back on track to achieve the plan objective.

X-Matrix

X-matrix is Hoshin Kanri's core process tool and strategic plan execution document containing all required critical plan information integrated comprehensively and synchronized across the organization.

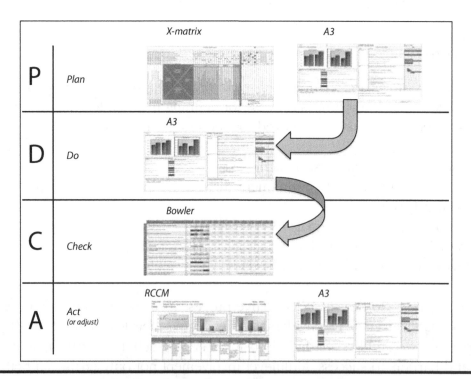

Figure 5.9 Lean enterprise planning and execution management.

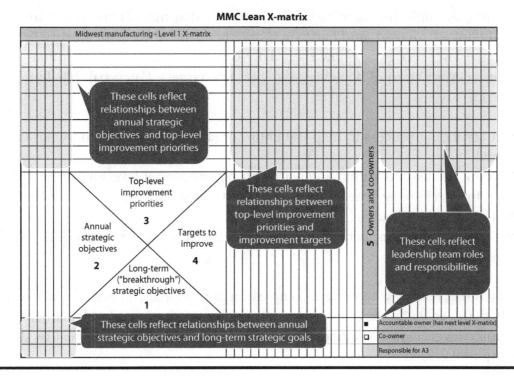

Figure 5.10   Lean enterprise X-matrix.

The critical information includes

- Long-term breakthrough strategic objectives
- Annual strategic objectives
- Top-level improvement priorities
- TTI's—Targets to Improve
- Consolidates long-term and short-term objectives on single page
- Communicates strategy relevance to daily management priorities

   X-matrix (Figure 5.10) captures an organization's annual plan very comprehensively in a single page. Specific content includes long-term breakthrough strategic objectives, annual strategic objectives, top-level improvement program priorities, and TTI. In addition to organizational alignment, X-matrix is a powerful communication tool providing clear prioritization guidance for an organization's daily activities and decisions.

Catch Ball

X-matrixes are established through a multi-level iterative process engaging all levels of leadership ensuring the plan is grounded in reality. It address required capabilities and resources in support of projects across an organization, resulting in organizational plan commitment. This process of cascading through the organization also ensures that metrics and goals are directly linked, aligned, and hardwired together. The result is that every organizational level recognizes how their daily operational activities contribute to and are a critical part of achieving the vital few targets. Catch ball has four elements:

- A scrubbing process to help the entire organization understand what is real.
- Frank discussions within and across all organizational levels: what prevents us from achieving True North? How can we best contribute?
- Resource level loading.
- Organizational engagement to gain commitment to achievement of the vital few goals.

Catch ball naturally engenders complete organizational commitment across all levels to the goals. In the scrubbing process, all organizational levels participate in frank conversations, resulting in a clear understanding of what is real by discussing topics such as

- What prevents us from achieving True North?
- How can we best contribute?
- What conditions must exist for goals to be achieved?
- What needs to be done so all necessary conditions are met?
- Confirming questions such as
  - Are goals understood?
  - Are goals achievable?
  - Are resources sufficient?

Catch ball's honest give and take will

- Increase probability of success in achieving organizational goals
- Gain complete commitment of all organizational levels
- Ensure all resources are fully deployed against organizational priorities and goals

Catch ball is fundamental to achieving an important Lean principle of full organizational engagement and is a vital communication process for ensuring real organizational alignment.

A3

A3 (see Figure 5.11) applied to Hoshi Kanri-defined projects is the core execution management document. It is the planning and execution method for achieving

Company XYZ A3 template title	Dept/Business:		Date:
	Team:		Proj. Leader:
Title:	**Recommendations:**		
Background:	What countermeasures are being proposed?		
What is the standard of performance?	How do you know the solution will eliminate the root cause?		
What is the history/context for this story?	What other countermeasures were considered?		
What is the problem?	What was the process for selecting the recommended solution?		
Why is this problem important?			
	Plan:		
	What will be done?		
	When will it be done?		
	Where will it be done?		
	Who will be responsible?		
Current situation:	Who is accountable?		
What is current performance vs standard vs perfection?			
What are the gaps?			
	Follow up:		
	How will desired results be confirmed and sustained?		
Goal:	What risks exist and what is being done about them?		
What is the desired performance?	What is the review schedule?		
What objectives are to be achieved?	How will the solution be standardized and audited?		
Analysis:			
What are the problem's root causes?			
What requirements and constraints need to considered?			
What alternatives need to be considered?			

Figure 5.11   A3 template.

X-Matrix priorities and TTI. This persuasive one-page living summary gets people to act in ways they otherwise would not by

■ Building a comprehensive situation analysis, requirements definition, and target conditions, which must exist to achieve each plan goal. Each plan is documented on an A3-sized sheet.

■ Engaging organizational members in constructive debate on what is required to achieve plan goals. This robust debate at all leadership team levels ensures that all conditions required for success are addressed. This reduces the likelihood that errant assumptions are made by leaders who may not have sufficient grasp of details to conditions "on the ground" and requirements that must be met to succeed.

■ Achieving target clarity, an understanding of what needs to be done, along with consensus and commitment of key project team members to effective and rapid implementation.

■ Reducing or eliminating lost time in meetings and discussions as they are focused on required conditions for success, not constantly debating plan assumptions and resource sufficiency.

When Is an A3 Required?

It is required for every top-level improvement priority target to improve and is initiated during completion of the X-matrix. Revisions needed during the year are made once there is consensus to make the revision, ensuring that each planned project achieves its established goals. A3s are a flexible and comprehensive tool that should be used to manage all complicated projects or problems.

Who Does the A3?

The A3 is prepared by project owners, and they are responsible for keeping it updated because of new learning, changes in situations, and progress made. Annual plan A3 documents are completed by project owners and discussed with their teams during the catch ball process. A3s are only of value when they are "living documents." Project owners are responsible for supplying updated A3 project documents for scheduled reviews and to apply countermeasures and PDCA when necessary to stay on schedule. An excellent resource to learn about and apply A3 is John Shook's book, *Managing to Learn*.[7]

Applying A3 within the Hoshin Kanri Process

Periodic Review Process, the C in PDCA—Bowler

Bowler is Hoshin Kanri's implementation scorecard. TTI are populated on a reporting format called "Bowler," which is used to track and manage every function and their department's TTI (Figures 5.12 and 5.13).

- The Bowler periodic review method is Hoshin Kanri's check step.
- The Bowler periodic review method tracks TTI against plan (red, yellow, green).
- Only off-track items are discussed and explained to determine their root causes, and the Bowler periodic review method implements countermeasures and/or solutions.

When is the Bowler used?

- During monthly reviews
- Core agenda of operating committee, functional leadership teams, and operational unit teams

Who owns the Bowler?

- The business operating committee, functional leadership teams, and operational unit teams own Bowlers established at various organizational leadership levels.
- Populating and updating are normally done by the project leader and team members responsible for the A3.

Populating the Bowler...TTI's from the X-matrix

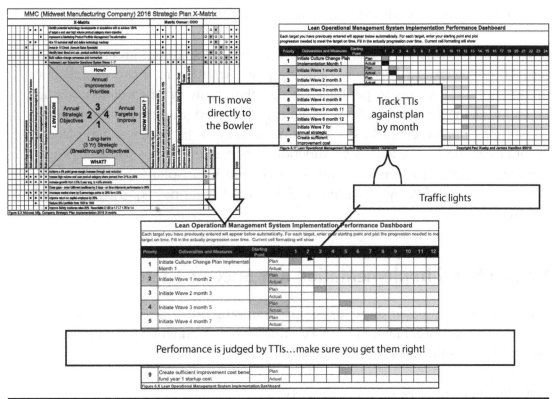

Figure 5.12 Populating the Bowler.

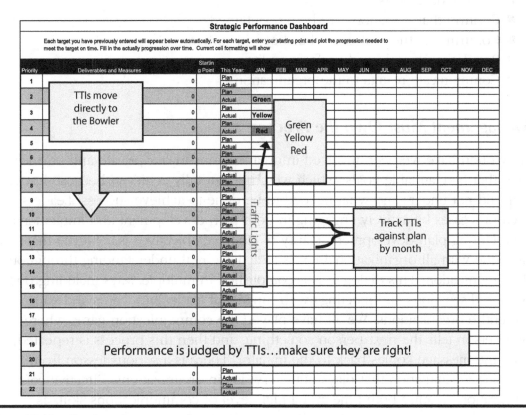

Figure 5.13 Bowler progress review.

The Bowler becomes the only form of dashboard used within an organization and must replace all existing dashboards. Once populated, the Bowler is used in regular operational reviews, tracking progress on TTI and strategic objectives. The reviews focus on areas with project status in green, yellow, or red. All reviewers participate in defining plan deviation root causes and developing countermeasures to get back on the plan. Only off-track items that are critical to achieving the vital few goals are discussed along with countermeasures being implemented to get back on track. Periodic reviews take place at all organization levels, across all functions and value streams. Bowler progress reviews are an essential element for Hoshin Kanri plan success as they ensure accountability by providing visibility of deviations requiring remediating actions.

Hoshin Kanri Implementation Key Success Factors

Success in any endeavor depends on leaders doing the necessary things to ensure that their organization is successful. Seven essential elements must be present for Hoshin Kanri to be successful in any organization:

- Management must own the process.
- Participation spans all management layers.
- Identify deviations.
- Confirm that deviations are being contained.
- Confirm that the root cause has been identified.
- Regularly schedule and perform audits.
- Communicate results to the organization.

Management Must Own the Process

How often have we been involved in organizations in our private and business lives that had laudable goals, which were never totally achieved because leadership did not understand what needed to be done to achieve success? Leaders own processes by actively engaging in both plan development and implementation. During plan development, clarity of strategic objectives is achieved through dialogue with all organizational levels to achieve real understanding throughout the organization. This dialogue is accomplished in multiple ways including one-on-one discussions and meetings with functional leadership teams and with the entire organizations. We all have played the communication game where one person tells the next person something, and then this process is repeated until the message comes back to the initiator. There is a serious lesson from this simple simulation that plays out every day in organizations. Effective communications require both time and a high level of communications skill, as each organizational level and function needs to hear a message relevant to their role and have an opportunity to dialogue, including challenging their leaders. Often, uncomfortable challenges from outspoken team members provide leaders with their greatest opportunity to connect with their organization ensuring that they "get it." They may not express their views with great clarity or tact, but, frequently with further discussion, one finds a valuable insight and a wrong or partially correct assumption. Leaders commonly fail because their assumptions turn out to be wrong. The only way to minimize making wrong assumptions in strategic plans is for leaders to dialogue with all levels accessing their knowledge and understanding of conditions that must be met for the goal to be achieved. Hoshin Kanri must be actively led, not simply delegated. Leaders lead by doing what they say they are going to do; it is the actions not words that stick in an organization's mind.

Participation Spans All Management Layers

Organizations operate on energy, the fuel propelling team members to overcome obstacles and achieve their objectives. One required leadership action is to build up an organization's energy level. This is accomplished through engagement during both planning and implementation. Catch ball ensures that conversations are meaningful. This ongoing process of dialogue and verification of mutual understanding is critical to an organization maintaining maximum energy. In Lean, we believe that the only people who truly add value in an organization are those directly involved with products and services valued by customers. This includes shop floor operators, sales, customer service, technical service, and product development. It is therefore obvious that the only change customers will experience will be delivered from changes made by those value-added team members. In addition, elimination of all waste is a primary Lean goal, so constant dialogue must occur through regular reviews, as projects and programs are adjusted to increase their likelihood of success. Hoshin Kanri ensures that constant and constructive dialogue is occurring throughout a business organization.

Identify and Ask the Right Questions

Our experience as leaders teaches us that asking the right questions is more important than having the right answers. It is through being asked the right questions that organizations begin to examine their preconceptions and assumptions. This engenders true understanding and knowledge, which leads to taking productive actions by resolving root causes of problems and/or creating the necessary conditions for success.

Deviation Requires Immediate Containment

We all use the phrase "stop the bleeding," and when deviations are detected, this is the first action that must be taken. This is true of shop floor operations or lost sales at customers. Organizations must have a sense of urgency when deviations occur; they cannot be viewed as just normal everyday business. Containment is step 1 in preventing further damage from lost productivity or customer dissatisfaction while rigorous root cause analysis is undertaken and corrective action is defined and implemented.

Schedule and Perform Audits Regularly according to the Plan

Maintaining schedules of regular reviews provides a venue for celebrating successes and also collectively addressing plan deviations. The old adage "success breeds success" is true, and recognition for successes shared by a team should be a standard part of every review. It is also important that plan deviations are addressed on a timely basis through regular reviews, an opportunity to engage the wider organization in assisting in both problem containment and determination of corrective actions to resolve root causes.

Communicate Results and Hold the Organization Accountable

Regular reviews should also be quickly followed by communications to the entire organization. Timely communication provides a great opportunity to reinforce organizational pride in achievements, energizing the team and increasing their commitment to more success. It is also important to communicate before the informal network broadcasts its version of the truth.

Organizational Deployment Approach and Model

The company's business and functional organizational structure needs to be considered when planning for implementation (see Figure 5.14), and there is not

"The Business Execution Process (Hoshin Kanri) in operations links together the strategic intent, operating plan and supporting programs and projects to drive results. In a very short period of time it has become a part of the fabric of division operations. My gauge of the success of a new approach is to think of whether you could go back to the 'old' way of doing things. In the case of BEP, I can't imagine going back. The BEP is now woven into our division operating rhythm. It is used at the Division Operating Committee, monthly Operations Reviews, and Top 20 Key Project Reviews. The natural focus is on the results and meetings are more effective."

3M Division Operations Director

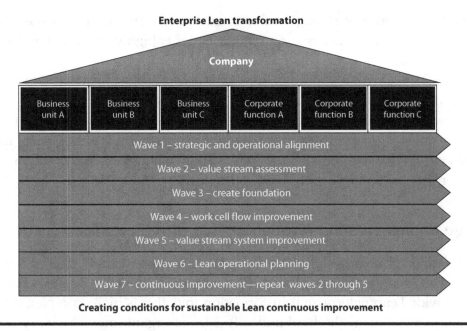

Figure 5.14 Lean enterprise transformation organization deployment.

a single approach that fits all businesses. There are a few principles to apply in choosing where to start.

1. Take a long-term strategic view regarding implementation. It is more important to build the LEOMS's culture on a solid foundation than to be overly preoccupied with getting the entire system implemented quickly. Building a strong and sustainable LEOMS is a multiyear process, and getting it done right is more important than getting it done fast.

2. The LEOMS is not strategic, but its implementation is strategic as it will improve a company's strategic market power. It is very important to introduce and communicate implementation strategic benefits so organizations like marketing and product development understand that this will help them be more successful in executing their organizational roles and not merely focused on operational efficiency.

3. Consider the company's market(s), financial situation, and effectiveness of its market discipline and differentiating competencies. Making improvement must improve a company's strategic position with customers first, followed by its internal financial improvement.

4. Clearly define and communicate the company's supply chain vision and operational model; while this may be obvious to supply chain professionals, it may not be to other organization members. Improvements must be made in a context of strengthening the supply chain model, not just reducing cost.

5. Make strategic choices regarding where to start implementation: If the company has multiple business units, ideally, one should start in the business

that has the greatest strategic value to customers. The most important criterion is selecting a business unit whose leader is highly respected and enthusiastic about taking the challenge of leading the initial implementation of the LEOMS. Obviously, if a business unit is in a severe financial crisis, this would likely drive the decision about where to start.

6. Be prepared to make organizational structure changes when a unit or department is unable to change their norms, practices, and behaviors to support and enable the LEOMS's timely implementation, collaboration with other groups, and/or achievement of improvement results.

Manage Implementation through an Executive Level Lean Leader

No one would consider building a bridge, road, or plant without a strong project leader and highly skilled and experienced project manager. Implementing a company wide Lean enterprise transformation initiative is more complex than building a new plant, as the hard and soft sides of organizations are very complex. It requires sufficient structure and skills to be successful. The Lean executive needs to be supported by strong project management skills that often are not inherent skills in many organizations except in specific groups that build plants, design and implement new process equipment, or implement IT systems. Success requires an overall Lean executive leader who has experience managing complex projects and deep Lean enterprise knowledgeable. It is important the Lean leadership position is a top executive level position to ensure easy access and credibility with company leaders and their organizations. The Lean executive leader needs to be supported by dedicated Lean specialists who are responsible for training and deployment across their assigned organization. There is no single organizational model that fits all businesses or projects, but there are universally required skills and leadership capabilities.

Training Process

Training Method

The training approach is another important success factor. It must be rigorous as building process discipline is fundamental to success (see Figure 5.15). All the training within industry documentation and processes are readily available for use by Lean coaches and is an excellent resource. The training method described is intended for both implementation of the LEOMS and building a continuous improvement culture. Figure 5.16 lists the training responsibilities of each organizational level. Lean coaches, area managers, and group and team leaders must be trained on the LEOMS, its implementation process, and tools prior to starting the first implementation assignment. This schedule covers the roles of all team

Training method

Roles and responsibilities:

Lean coach and practitioners to conduct regular training / knowledge transfer at plant sites and identify tasks to be completed each month.

Train – Apply – Practice – Verify – Review

Lean team leaders will maintain a list of planned actions for the month and maintain over-all schedule/activities in their area of responsibility.

Lean coaches will share knowledge from all plants to accelerate learning and internalize Lean knowledge and implementation skills.

Lean coach at each facility is responsible for leading the implementation project including coordinating and training project teams. This coach will work with additional site resources as required to achieve the implementation schedule.

Plant management at each site will be the owner of the Lean Operational Management System transformation implementation schedule

Figure 5.15   Lean implementation training method.

Roles and responsibilities by wave – Realizing Ninjutsu

Train – Apply – Practice – Verify – Coach

	Strategy, alignment and launch	Resource, train, and assess	Create foundation	Work cell redesign	System deployment	Sustain continuous improvement	Annual plan deployment
Team members			✓ Lean orientation ✓ Lean 5S training ✓ 5S participants	✓ Lean training ✓ Project participants ✓ Kaizen participants ✓ Initiate improvement ideas	✓ Lean training ✓ Project participants ✓ Kaizen participants ✓ Initiate improvement ideas	✓ Recognize waste ✓ Project participants ✓ Initiate improvement ideas ✓ Kaizen practitioner	✓ Recognize waste ✓ Kaizen practitioner ✓ Support the team
Team and group leaders		✓ Lean orientation ✓ Lean system training ✓ VSM assessment participation	✓ Lean training ✓ 5S participants ✓ Daily audits ✓ Coach	✓ Lean training ✓ Project participants ✓ Kaizen participants ✓ Daily audits ✓ Coach	✓ Lean training ✓ Project participant ✓ Kaizen participant ✓ "Evangelist" ✓ Daily audits ✓ Coach	✓ Lean training ✓ Project participant ✓ Kaizen participant ✓ "Evangelist" ✓ Daily audits ✓ Coach	✓ Project leader ✓ Kaizen leader ✓ "Evangelist" ✓ Yokoten facilitator
Lean coaches/ area managers	✓ Select and train Lean coaches	✓ Coach training ✓ VSM assessment leader ✓ Be the evangelist ✓ Facilitate Yokoten	✓ Coach training ✓ 5S launch ✓ Knowledge transfer ✓ Facilitate Yokoten ✓ Scheduled audits	✓ Coach training ✓ Project leader ✓ Knowledge transfer ✓ Facilitate Yokoten ✓ Scheduled audits	✓ Coach training ✓ Kaizen and project leader ✓ Knowledge transfer ✓ Facilitate Yokoten ✓ Scheduled audits	✓ Coach training ✓ Kaizen and project leader ✓ "Evangelist" ✓ Knowledge transfer ✓ Facilitate Yokoten ✓ Scheduled audits	✓ Train and develop project teams and employees ✓ Lean coaching ✓ Facilitate Yokoten ✓ Scheduled audits
Leadership	✓ Align business strategy and Lean ✓ Communicate case for change ✓ Implementation timeline ✓ Culture change plan ✓ Commit resources	✓ Executive training ✓ Apply Hoshi Kanri ✓ Prioritize resources ✓ Critical few metrics ✓ Bowler progress tracking	✓ Executive training ✓ Bowler progress tracking ✓ PDCA ✓ Drive Yokoten across facility operations ✓ Shop floor visibility	✓ Executive training ✓ Bowler progress tracking ✓ PDCA ✓ Drive Yokoten across facility operations ✓ Shop floor visibility ✓ Communicate successes	✓ Executive training ✓ Bowler progress tracking ✓ PDCA ✓ Drive Yokoten across facility operations ✓ Shop floor audits and coaching ✓ Communicate successes	✓ Quarterly executive Lean review ✓ Bowler progress tracking ✓ Drive collaboration in the business ✓ Shop floor audits and coaching ✓ Communicate successes ✓ Drive Yokoten across facility operations	✓ Integrate Hoshin-Kanri into annual planning process ✓ Host coaches quarterly lesson's learned ✓ Company wide communication ✓ Shop floor visibility ✓ Lead Yokoten across facility operations
Lean knowledge partners	✓ Train ✓ Lead first application ✓ Assess and coach leaders and teams	✓ Train ✓ Lead first application ✓ Assess and coach leaders and teams	✓ Train ✓ Lead first application ✓ Coach and assess leaders and teams	✓ Train ✓ Lead first application ✓ Assess and coach leaders and teams	✓ Train ✓ Lead first application ✓ Assess and coach leaders and teams	✓ Train ✓ Lead first application ✓ Assess and coach leaders and teams	✓ Train ✓ Lead first application ✓ Assess and coach leaders and teams

Figure 5.16   Roles and responsibilities by wave.

members and leaders during the initial LEOMS facility operations implementation and ongoing sustainment:

- Train means to become knowledgeable about the assigned processes:
  - Review all related documentation including all current standard work procedures, production reporting, connecting activities, process measurement, and quality inspection that are part of the process's standard work documentation.
  - Walk the process from the last step back to the first step, reading the process documentation at each step.
  - Talk with team leaders, front line team members, maintenance technicians, and engineers as their experiences and perspectives are all valuable, and they will all be important resources to achieve success in improving operations and sustaining continuous improvement.
- Apply means to show frontline team member trainees the process steps and all related activities and have them do the process until they feel comfortable and they have grasped all the activities.
- Practice means to have trainees do the process, following the standard work with the coach observing. This should be done until the trainees demonstrate that they have command of the standard work steps.
- Verify means trainers should return to verify that frontline team member trainees are preforming assignments in alignment with standard work instructions over a number of cycles. Trainees should be left on their own to perform their assigned process, and coaches should periodically return to verify that the job is being done per the standard work instructions. This should be done until at least three consecutive verifications show that trainees have mastered the process. At this point, the regular verification will be done according to standard work verification responsibilities assigned to the group/team leader and area manager.
- Audit means team leaders, group leaders, and area managers complete scheduled audits in compliance with the audit schedule.
- Coach means to provide encouragement and feedback to assist operators in improving their daily performance; as operators become proficient in doing the assigned task, it is also important to teach them the "whys," providing rational explanation regarding their assigned work tasks.

Audit and Review

A review schedule should be implemented to review standard work procedures with assigned operators both to reinforce its application and also to stimulate improvement suggestions. It should be noted that Lean Coachs' training process role only applies to the system implementation phase as Area Managers will have responsibility for daily operations audits and training of Frontline Team Members

> ## LEADERS MUST GET RESULTS, BUT THEY ARE ALSO OBLIGATED TO BE BUILDERS
>
> Thirty years ago, I was a plant manager and my daughter gave me a framed quotation from the old testament, Proverbs 29:18 *Where there is no vision, the people perish.* I have had this on the front of my desk as I moved through nine assignments since then. It reminds me every day, that operational performance is a minimum standard, providing a clear vision to which the organization should aspire and by which it should be challenged, is a leadership obligation. We are fortunate to be given the opportunity to lead; we need to be sure that we are good stewards of what we are given and leave things for the next generation better than we found them.
>
> **Paul Husby**
> *Leaders*

once system implementation is complete, verified by the assigned coach and confirmed by team leaders and group leaders. Coaches are responsible for training new plant team members and doing "random" audits with shop floor leaders to maintain consistency and manage new system changes. The standard for making this transfer is three consecutive successful verification audits, as this demonstrates competency in completing all assigned work cell tasks. The second training step is teaching PDCA to frontline team members once process operations ownership has been taken by frontline team members and group and team leaders. Lean coaches are responsible for continuing to improve the PDCA skills of team leaders, group leaders, and frontline team members.

Review and Finalize the Business Case

With culture changes defined and Lean enterprise better understood, it is time to create the business case for becoming a Lean enterprise, "the compelling why"; a company's rational for choosing to become a Lean enterprise articulates the importance of and need for making this choice. The business case communications must be designed to be relevant to all constituencies, customers, team members, the relevant communities, and investors as it is a very big change initiative. The business case should commit to both near-term (1–3 years) and longer-term (5–10 years) goals reflecting what will be different about the company and what market and financial benefits will be achieved. There must be a compelling cause strong enough for organizations to embrace Lean's transformational change. The first prerequisite of significant change is motivation; people must have a cause justifying their commitment, energy, creativity, and willingness to change behaviors and practices. Both authors have had experience with businesses that failed or were threatened with failure and those that were in a "comfortable" market and financial position. We all are aware of the expression *the fear of death can be an incredible motivating force.* Many companies are not under immediate threat of going out of business. So how do they create their cause to inspire total organizational commitment and energy to make significant change?

The key to creating a cause is in radically changing the lens through which organizations see themselves. There is great strength in changing magnification power to view a wider range of possibilities, a new scale of comparison. Define and review the trends occurring in business globally and what has happened in the past to companies who could not or would not change to maintain their competitiveness. The entire competitive field must be viewed while considering long-term threats and sustainability (remember the company survival rate data presented in Chapter 2). In short, change the organization's perspective about itself and its current levels of performance to increase the odds of long-term success and survival. It is important to align all team members' interests, such as sustaining good pay, career growth, a healthy work environment, and job security as benefits of becoming more competitive by adopting the LEOMS.

What Is the Compelling Vision? Formulating and Communicating the Cause

In his book *Confronting Reality*,[8] Larry Bossidy states, "The tools, practices, and behaviors that will distinguish success from failure can be summed up in one phrase: relentless realism." Frequently, leaders with good intentions fail to confront reality because they do not want to take the required risks of transformational change. This process always involves organizational uncertainty, potentially affecting people who have been colleagues for many years. The leader also assumes significant career risk because of potential failure or board of director impatience. Another common reality is avoidance behavior resulting from "arrogance" brought on by past success. Companies with strong intellectual property or with superior market positions frequently believe that "we're different," "we're better," or "that won't happen to us." Unfortunately, avoiding reality leads to the worst outcomes for leaders, companies, and its employees because it opens opportunities for aggressive competitors. Leaders are responsible for keeping organizations grounded in reality because competitive advantage is temporary and must be continuously rebuilt and maintained. Most people want to know exactly where they stand and feel respect for a leader who keeps them grounded in reality. Executives who shield people from reality and its risk also deny them an opportunity to make choices involving their career, family, and the company. Doing this underestimates people, not believing in their ability to make the right choices.

Organizations are motivated to embrace change when its members are fully informed about the risks to success of their company as they feel that leadership respects them as full partners in company success. Toyota's leadership demonstrated this in 2002 when they asked all plants and suppliers to reduce costs by 30% to 40% in 3 years to counter emerging threats from Korean and Chinese auto manufacturers. By 2004, the Georgetown Kentucky engine plant was well on its way toward achieving these goals, a clear demonstration of Toyota's forward thinking and the LEOMS's continuous improvement power. While Toyota more recently has tripped over their feet of clay, they are doubling down on getting back to living the "Toyota way." Competitors may take some short-term satisfaction in Toyota's failures, but one should not bet against their previous five decades of consistent pursuit of perfection. Most businesses, like Toyota, are not on the verge of failure, but how many can engender commitment to make substantial change when business is good? It is done by leaders who are able to inspire their organization to truly commit to reaching for the best both as a business and as professionals.

Creating Sustainable Competitive Advantage—Ten Steps to Success

Long-term success in any business depends on having differentiation in one of three market disciplines: product leadership, customer intimacy, or operational excellence that customers value enough to be long-term purchasers of their product or service. This market differentiation is always built on differentiating competencies and great operational execution.

Step 1. Commit to Investing in Startup Resources

Big change initiatives are never free! When properly implemented, the LEOMS will cover implementation project resource investments in the first year or two, and begin to generate improved profit margins and increased cash in year 2 or 3. These improvements create resources to invest in growth opportunities and new technologies without reducing gross margins. The initial investments include hiring an experienced corporate Lean leader and full-time Lean coaches. The corporate Lean leader should report to the president or CEO as Lean involves all functions and not just supply chain operation, as it is important to avoid internal "political subversion." Placing responsibility too low in the organization exposes Lean enterprise implementation to the "invisible hand or hands" of leaders who subvert the implementation for their own personal reasons, often involving excessive personal pride as Lean can feel like a criticism of the past. The leader's role is to select a Lean coach for each business unit executive and be responsible for training the organization and driving implementation organization wide.

Implementation focuses on both bottom-up and top-down improvements. The bottom-up focus is in training all operations team members and their leaders from executives to plant managers who must have an understanding of the LEOMS and be prepared to invest time and resources in each operational facility to train their organization and initiate a first wave of improvements. This is critical to ensure that momentum is built as organizations see improvements and take ownership for applying the LEOMS to their operations. Simultaneously training business and functional executives along with business planning to adopt Hoshin Kanri will have immediate impact on getting all the organizations on the same page and improving clarity and focus on improving business plan execution.

THE CHALLENGES OF BUILDING THE ORGANIZATION STRUCTURE TO SUPPORT DAILY KAIZEN

Despite the benefits of daily kaizen, one of the main reasons that daily kaizens interrupted or stopped is that from a financial standpoint, it can be hard to justify allowing individuals to spend 10, 20 or 30 minutes several times each week engaged in identifying deviations and problem solving activities. The immediate attention to costs must be balanced by the larger goal of constructing the new culture norm of problem solving everyday. Often an opportunity to help offset this cost is utilizing the productivity benefits of redesigning existing material flow and work station redesign using Lean tools and methods, instead of reducing headcount as a result, investing in the creation of the group leader position that will speed up problem solving learning and cover the operators while they are working on dailyproblem solving.

Adapted from Creating a Kaizen culture,[9] p. 131

Step 2. Manage the Plan

Creating a vision using benchmarking with best-in-market companies is useful to establish improvement goals and to build commitment to LEOMS implementation by applying Hoshin Kanri. The LEOMS's implementation project leaders must establish project plans to meet performance improvement goals, identify significant stages of progress or milestones, and manage implementation using the Bowler. These project milestones are the focus of monthly project reviews. Frequently, project leaders spend hours completing detailed project plans with all activities using project-planning software. They often focus on activity tracking and rescheduling, assuming this is project management. Unfortunately, this common practice usually leads to failure. Lean coaches must do several things for their Lean enterprise project to succeed using an A3 (see Figure 5.17)

■ Define the project its scope and deliverables
■ Establish and track project milestones

| Midwest Manufacturing Company | Dept/Business: Human Resources | Date: |
| | Team: Operating Committee | Proj. Leader: H.R V. P. |

Title: Company Cultural Assessment	**Recommendations:**
Background: Only 15% of major improvement initiatives are successful in meeting the stated objectives. The major reason for failure is the lack of cultural compatibility with cultural requirements for success of the Lean Enterprise Operational Management System.	Recommendations will be made to change all cultural attributes that are not compatible the Lean Enterprise Operational Mangement System.
	Plan: The plan will make recommendations regarding what if any culture changes must be made and changes to processes, performance appraisals, and behavioral norms.
Current situation: Midwest Manufacturing Company is a family owned business founded in 1950 that supplies sensors and measuring products to a wide variety of manufacturers. Recently with intense competition from imported projects company growth has slowed to 2% annually. Historically they have relied on dealers, distributors and manufacturers' representatives to manage end customers and have assessed the need to have more direct presence in their major end user customer.	
Goal: Determine what if any culture changes are needed at the company, staff and functional organizational. This will be determined by applying the Organzational Culture Assessment Instrument to team members at all levels of these organizations.	**Follow up:** The follow up will continue for years to insure changes made are sustained and in fact embedded in a new culture compatible with the Lean Enterprise Operational Management System.
Analysis: The analysis will involve using the cultural assessment of each organization at every level to determine what if any cultural attributes are not compatible with success of the Lean Enterprise Operational Management System.	

Figure 5.17   A3 MMC culture change.

- Understand project risk related to each milestone and have contingencies prepared if failure appears probable
- Understand milestone critical paths, which cannot be off schedule, or which projects are most at risk of not achieving their goal

Maintain all projects using the Bowler to conduct regular monthly project reviews. Executive leaders do not need to be project experts, but they must question their Lean coach about the A3 processes and project progress. Only constant follow-up will ensure expected results and early detection of potential failures while time is still available to get back on schedule.

Step 3. Anticipate and Plan How You Will Overcome Resistance to Change

This is especially important when viewing how to manage the "three tribes" always present in an organization in the process of change. There are the early

3M SIX SIGMA RESISTANCE

Six Sigma at 3M was embraced by the entire organization with one exception, the 3M Research and Development Laboratories. The failure occurred because company top leadership and Six Sigma program leadership failed to recognize the laboratory culture's strong adhocracy and clan culture attributes that resisted the "rigidity" and focus of Six Sigma on operational process improvement to reduce cost. This could have been avoided by recognizing these unique culture attributes and focusing laboratories on applying DFSS, Design for Six Sigma, that starts with establishing customer requirements to focus product design and on meeting those requirements through manufacturing process design supporting Six Sigma quality levels. The unfortunate outcome was a "below the surface" rebellion against Six Sigma and an organization that was unable to embrace and appreciate the contributions Jim McNerney made to 3M.

Paul Husby

adopters, hardcore resistors, and the observers, so it is important that early adopters feel confident about promoting change and give evidence to observers to mitigate the negative influence of hardcore resistors. Normally, the largest group will be observers, but as they begin to choose to get on board, pressure increases on hardcore resistors.

The goal is to turn observers into believers, and applying the force field analysis is an excellent tool for figuring out what to do (see Figure 5.18). Observers are

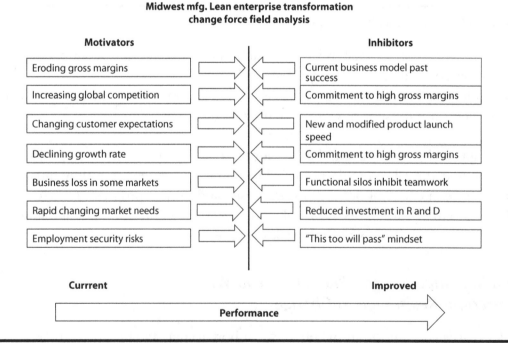

Figure 5.18 Change force field analysis.

normally honest skeptics who need to get comfortable with proposed change so information needs to be gathered from the observers to identify the motivating forces in favor of change and inhibiting forces preventing observers from getting on board. This is not to say that both early adopters and hardcore resistors should be ignored, as they have important inputs that need to be represented in the force field analysis. This information facilitates creating a plan to reduce hardcore resistor influence and increase the motivating force of early adopters. Developing the plan involves creating credible logic and arguments to persuade observers to join the cause; for example

- Look back at company history to people and events demonstrating that overcoming big market changes is part of company history and broadly communicate these stories
- Increase resources in research and development to increase innovation resulting in new products and improved current products
- Define growth market segments that will be focused on reviving company growth
- Communicate basic research and technology programs focused on future market needs
- Become a Lean enterprise to create resources to reinvest in growth initiatives
- Discuss the culture change plan and speak directly about the consequences of not changing

Working the plan will lead to observers getting off the fence and joining the early adaptors. Once there is a critical mass of observers on board, the hardcore resistors will see the writing on the wall and either choose to get on board or leave the organization; those that stay will eventually evolve into the strongest vocal supporters of change. So, do not treat the vocal resistors as adversaries; instead, see them as future potential strong leaders who will be important to achieving success.

Step 4. Develop a Communication Plan

Communication keeps organizations informed, engaged, and aligned. A structured communications process is a valuable element of transformational change, as it

1. Maintains a disciplined schedule for regular communications
2. Keeps team members informed on the state of the business
3. Communicates improvement process successes and failures
4. Is an opportunity to recognize team members for their contributions
5. Gives team members an opportunity to question leadership and be heard
6. Provides an external view of the company from customers and investors

Communication is leadership's main tool for transformational change. Confusion, misunderstanding, anger, disagreement, and suspicion will all be present in an organization involved in transformational change. Regular communication from every leadership level is required to acknowledge people's feelings and at the same time reinforce the bigger cause for change requiring continued commitment. Ninety percent of team members will accept and live with things they do not completely agree with as long as they have had an opportunity to be heard and feel respected. Allowing people to keep their own feeling about change also enables them to change their opinion when they see growing company success. Nothing prevents people from changing their point of view more than resentment and anger generated from feeling that they have been disrespected and not heard.

- Transformational change is about building a wave of positive organizational support, which gradually overcomes resistance of all but the last 5% who may never get on board. This last 5% will be resolved as they either get on board, decide to leave, or are asked to leave because they are no longer aligned with the organization culture, values, and norms.
- A formal communications plan has four elements:
  - Vision communication by the CEO to the entire organization
  - Cascading communications throughout an entire organization to make sure everyone understands key messages and to confirm every team member will be provided training as a part of implementation
  - Quarterly organization-wide communication meetings and periodic assessments by human resources and the communications department to provide feedback on leadership's performance and organizational understanding, acceptance, and commitment
  - Communication plan content
    - Business case for change
    - The expected company benefits
    - How change will affect team members
    - Culture change plan, rationale, corporate, and function culture changes
    - Lean Operational Management System implementation timeline

Step 5. Launch Lean Enterprise

LEOMS's Launch Readiness Assessment

The Launch Readiness has two valuable applications in getting the LEOMS implementation off and running on a path to success. The checklist (see Figure 5.19) serves as a planning checklist to be sure that all the necessary planning, preparation, training, and resources are in place to get off to a fast start along with being an objective way to assess progress and take any needed corrective actions ensuring that launch is successful.

Lean enterprise implementation readiness assessment	Needs work	Acceptable	Ready to roll
Ready to roll - 89 points or greater - keep working on the 3's to move all to 5's Work <89 continue learning and development moving 1's to 3's and 3's to 5's			
Item with a grey box must be at readiness level 3 or 5 as noted	1	3	5
1 A vision, mission, guiding principles and business case to support Lean enterprise implementation exist			▓
2 Cost, cash and growth targets are established*			▓
3 Culture assessment completed and change plan in place			▓
4 Lean operational system deployment rollout plan is defined and reviewed by top leadership			▓
5 Top leadeship team is trained and motivated to be visible champions of the transformation			▓
6 The implementation team is trained and ready to lead the implementation			▓
7 Lean system deployment projects are supported by credible project scopes and resources			▓
8 Sufficient outside expertise has been contracted to support the Lean enterprise deployment plan			▓
9 The entire organzation has been engaged in two way coversations about the proposed changes			▓
10 Top talent has been selected and trained as Lean coaches			▓
11 An implementation plan including key milestones is in place			▓
12 People in organizations directly involved with implementation start-up are trained and prepared to start			▓
13 A force field analysis has been completed and a plan is in place to overcome identified resistance			▓
14 Accounting charged with calculating total department actual vs target project benefits**		▓	
15 Culture changes expectations are included in performance reviews with quarterly feedback		▓	
16 Sufficient dedicated resources are available to support the implementation in all functions		▓	
17 All the organization has been appropriately communicated with based on how their position is impacted		▓	
18 Leaders are committed to regular listening sessions and individual conversations to gain feedback		▓	
19 Regular reviews are scheduled to recognize and celebrate progress		▓	
20 HR policy includes consideration of Lean knowledge, coaching skills and results in all promotions		▓	
21 Improvement expectations integrated into every operational review;		▓	
Assessment total =			
*Measurable hard savings against leadership targets; soft savings should count only when directly connected to future strategic goals			
**After one year a percentage of the expected benefits (30–50 percent will be factored into P&L's and operating budgets)			

Figure 5.19 Lean implementation readiness assessment.

Step 6. Selecting and Training the Implementation Teams

Every business unit and corporate function must have a Lean coach reporting to its top executive and must have an appropriate number of trained Lean coaches that are sufficient to provide the organization with the LEOMS training and support. These top business units and functional Lean coaches will be responsible along with their executives for implementation in their units. The culture change plan and implementation should be led by human resources and business and functional executives. The LEOMS's training must have a classroom component, a hands-on practical application learning lab, and coach-guided implementation review sessions. There are two high-level stages to implementing the LEOMS; first is constructing a well-designed system and putting it in place. The second stage is the hard work of training the entire team and coaching them to become competent at applying kaizen daily. This will be covered in more detail in Chapter 6.

Step 7. Make Lean Enterprise a Company Wide Reality

Implementation speed will depend on management's prioritization of sufficient Lean coach resources to deploy the system. Some thought must be given to selecting the first implementation site or organization, one should consider the organization's internal credibility, past success with making changes, and its leaders' enthusiasm for implementing the LEOMS in their plant. The idea of pilots should be discarded, as the LEOMS is proven over decades, so the only question is the commitment to make it happen. Secondly, continuous improvement is a fundamental part of the LEOMS, so learning by doing is the best solution; if good results are not obtained, the root cause is not going to be the LEOMS.

Step 8. Build and Maintain Momentum by Achieving Early Results

Nothing provides energy, credibility, and confidence to an organization like success. Early benefits are important criteria in initial project selection because they provide organizational reinforcement (see Figure 5.20). Early success is crucial to overcoming resistance, convincing uncommitted observers to get on board, thus isolating hardcore change resistors. The LEOMS can accomplish this by getting the system designed and up and operating in a value stream to bring immediate visible operational improvement and cost benefits. Lean coaches and functional managers are responsible for capturing and publishing success stories to assist in building momentum. Building participation is vital to long-term success, and Masao Nemoto (see Figure 5.21) provides his five principles for building participation. He was the managing director of Toyota Motors Engineering and Quality in the 1970s and 1980s; he was highly praised by Soichiro Toyota in the introduction of his book, *Total Quality Control for Management*.[10]

Step 9. Relentless Pursuit of the End State

Organizations are frequently critical of their leadership for focusing only on the short term. This has become a greater leadership challenge as investment analysts push for continuous quarter-by-quarter improvement in financial performance. Leaders must continually remind their organization of the reasons why

- Encourage reporting, capture and publication of success stories
- Top local leadership must maintain an active presence where the action is to provide encouragement and reinforcement of shop floor team members' development and contributions – it is the essence of building positive energy
- Spontaneous celebrations when a team member makes an extraordinary effort to assist another team member
- Team leaders should recognize team members at the start of shift meeting for positive acts the previous day
- The top department leader should meet with team leaders and area leaders weekly to answer questions and give encouragement
- Top company leadership must have a regular scheduled presence with all facilities, area sales organizations and corporate functional groups to reinforce the importance of the Lean Enterprise transformation

Figure 5.20 Start-up accelerators. Lean management coaching responsibilities—Lean Institute Executive Forum.

Principle I: Leaders must also be engaged in Kaizen
- Sets an example and reinforces leaders' credibility
- Leaders need to talk about improvement constantly
- Unless a leader talks about improvement after improvement, subordinates will have no incentive to consider improvement

Principle II: Show interest in subordinates' improvement activities
- What kind of improvement have you been able to bring to your workplace?
- What are you planning now?
- Respond with compliments regarding steps in the kaizen process and completed improvements. For example, compliment people for detecting a defect.
- Listen carefully to what subordinates have to say so they feel what they are doing is valuable and important
- Meet not only with large groups of employees but small groups on the shop floor around A3's or Poster board summaries of improvement projects

Principle III: Never say: What? How come you are doing that kind of improvement now?
- Do not criticize, this embarrasses and discourages subordinates from making improvements

Principle IV: Seeds of improvement are limitless
- Human beings have the ability for continuous improvement
- Making improvements enhances a person' ability to make more improvements
- Listening to examples of other people's improvements increases knowledge to make improvements
- Studying equipment, tools and materials results in new knowledge leading to more improvements

Principle V: Have ears to listen to the mistakes committed by subordinates
- Mistakes are seeds for improvement, take advantage of them to improve, don't criticize mistakes
- Listening very intently to demonstrate what the subordinate has to say is important
- Encourage the person who makes a mistake to see it as a seed for improvement

Figure 5.21 Coaching employee Kaizen participation. (From Masao Nemoto, *Total Quality Control for Management*, Prentice Hall, 1987.)

continuous improvement achievement of short-term benefits is an important first step toward achieving the long-term visions. Leaders must model a focused commitment to end state achievement. This builds their credibility for dealing with inevitable unexpected events requiring short-term reprioritization of projects to sustain overall improvement. These events will be viewed in their proper perspective once an organization has experienced sustained leadership commitment to their transformational end state LEOMS's journey.

Step 10. Conduct Regular Operational Reviews at All Levels

This review process must occur at all levels of the organization with greater frequency as it moves from top-level operational performance reviews to frontline team members. Daily frontline reviews are essential for maintaining operational discipline, continuous improvement, and employee development. Lean, for example, is a people- and leadership-based process with layered audit control plans. Operators, floor supervisors, and area managers have a daily standardized work routine to audit certain areas.

Plant managers spend their first hour or two each day reviewing frontline operations to audit, coach, and identify next areas of improvement.

- *Audit*: Leaders at all levels audit an area others have audited, ensuring real accountability because everyone knows that at some point their work is inspected. Audits are a preventive act, an early detection system identifying small deviations that may cause quality, service, and safety or cost failure if not corrected.

■ *Coach*: Lean teaches practitioners to see with *new eyes* and constantly pursue continuous improvement to achieve perfection. Tools and methods obviously assist in this process, but a greatly increased sensitivity to waste and its sources changes managers' thinking. This skill needs to be coached and nurtured everyday by questioning, for example, why the hourly production target was not achieved, and why something is done in a particular way. Why, why, why, why? This drive to understand, identify root causes, and implement solutions is also an opportunity for leadership to use their knowledge and experience to coach and mentor their team members.

CREATING ONE PLANT ONE TEAM

Before I became a plant manager I was the plant superintendent and my plant manager started the practice of all salaried people spending time on the shop floor a couple of times a year. The factory operated a 4 crew 24/7 rotating shift schedule so it was easy for a big gap to occur between the salaried staff and the team members on the crews. When I became plant manager and began to lead our plant in the implementation of Lean, we instituted a new practice of running executive tours. The plant operating committee became the tour guides and first level supervisors and crew members made all the presentations on the shop floor. I admit that the first time or two I was quite anxious as I didn't know exactly how our people would do and what the reaction would be from 3M top management. In the end it was a win, win, win! Our people showed pride in what they were doing and problems they were solving, I saw the ownership and commitment in our shop floor team and top management was quite pleased. In fact, the Chairman and CEO at the time had 3M's fleet of 6 Gulfstream G4's fly every executive to our plant for a tour. We hosted one group in the morning and another in the afternoon. I was pleased but the positive message this sent to our entire plant team was invaluable.

Paul Husby
Hosting the 3M Executives

■ *Identify next levels of improvement*: In addition to auditing and coaching, time must be spent observing operations not only to understand the details of a particular process but also to see how the entire system is functioning. This experience reinforces something that most leaders know but do not do, which is that time dedicated to operational reviews is critical to both process improvement progress and leadership development. Many times reviews turn into a search for the guilty and punishment of the innocent rather than coaching participants by asking questions about the improvement

Key change communications content:

✓ Identify the business issues that made the changes necessary

✓ How the change addresses these issues

✓ Acknowledge upsides and downsides of the change

✓ Give as much overall information as possible

✓ Offer milestones instead of hard deadlines

✓ Avoid cheerleading

✓ Have a follow-up plan for announcements

✓ Show and have interest in the their questions and concerns

Figure 5.22   Strategic communications tips.

opportunities, what problems they are working, what root cause(s) they have identified, and what countermeasures they have tried to resolve the problem. Leaders frequently want to show their knowledge, but instead demonstrate their arrogance and ignorance about their operations. Leaders must constantly remind themselves that they must also be learners and teachers.

Launching and communicating the LEOMS transformation plan, including the culture change plan, needs to be well planned and rehearsed. Figure 5.22 describes a valuable set of suggestions to guide introductory and ongoing communications related to LEOMS transformation, as one only gets one chance for a successful launch. The bottom line is that the LEOMS is not easy and must be approached and owned by the implementing organization as they learn the principles, methods, and tools along the implementation journey.

There is no short cut as "just copying" or reengineering will be a failure as pointed out by Mike Rother in the quote below.

> ### REVERSE ENGINEERING TO BECOME A LEAN ENTERPRISE
>
> Reverse engineering does not make an organization adaptive and continuously improving.
>
> **Mike Rother**
> *Toyota Kata, McGraw Hill, p. 9, 2010*

Wave 1: Strategic Alignment

Hoshin Kanri is applied to assure alignment of strategic and operational plan goals, hardwiring organizational integration and ownership, assuring a high probability of sustainable LEOMS implementation success. Wave 1 prepares organizations to successfully launch the LEOMS transformation and get wave 2 off to a fast start. The obvious starting point is articulating a company's value proposition and differentiating competencies and to make its business strategy explicit. This initiates organizational alignment by focusing on the LEOMS to drive continuous improvement of a company's value proposition, differentiating competencies, and operational performance, ensuring that improvement efforts will result in increasing customer value and the company's financial return.

Launching the LEOMS Implementation

The business strategic plan with 3-year targets and historical market, operational, and financial information creates the baseline and direction. The key input is the 3-year strategic and operational plan that addresses all the deliverables listed in wave 1: strategic alignment (see Figure 5.23). It is important to consider the two items in the first deliverables, value proposition and differentiating competencies, along with a clear understanding of the supply chain model supporting the delivery of the value proposition. These are areas where process improvement goals should be focused on increasing competencies, capabilities, and speed as they are the heart of a company's competitive advantage. Adoption of the LEOMS must be implemented in support of increasing customer value supplied by a company's value proposition and its supply chain model. It is counterproductive to just make changes that may have some factory cost reduction but may detract from effective value proposition execution. The goal is to continuously improve customer value, and that means improving customer experiences through

Strategic alignment deliverables:
- Company's value proposition and differentiating competencies
- Supply chain strategy – define based on demand profiles, volume seasonality, service strategy and model – make to order, make to stock or a hybrid
- Business process segmentation – strategic, operational, infrastructure
- Service requirements for key market segments and key customers
- Organizational deployment model and approach
- Three year capacity plan
- Inventory turnover goals by class (raw, semi-finished, and finished goods)
- Factory cost percent goals by product family and total
- Unit cost goals (at least for all finished product "A" items)
- Product performance and quality requirements
- Critical few metrics' performance Bowler charts – red-green-yellow charts
- LEOMS implementation business case and deployment plan with goals, milestones, and resource requirements
- Lean enterprise transformation communication plan
- Lean enterprise transformation launch, including culture change plan

Information required:
- Current strategic and operational plan outputs
- Current 3 year financial plan and history
- Product demand forecast and ABC analysis
- Market segment, sales, growth, market share and key customer service history
- Factory cost, unit cost, service and inventory metric history

Note: all goals and targets are for a 3-year time horizon unless defined otherwise

Figure 5.23   Wave 1: strategic alignment.

increased performance, capabilities, and execution of supply chain operations and strategic differentiating competencies.

MMC Hardwiring Strategic and Operational Plan TTI's

Application of Hoshin Kanri is a practical, easy-to-understand, and effective tool set for managing deployment, communication, and implementation of strategic and operational business plan and the LEOMS project implementation. Hoshin Kanri processes and tools will be applied to MMC's LEOMS implementation throughout the remainder of this chapter and Chapter 6. The starting point is MMC's 2016 Strategic Plan X-Matrix (see Figure 5.24),

Number 1: 3-year what?—Long-term strategic break through objectives
Number 2: How far?—Annual strategic objectives
Number 3: How?—Annual improvement priorities
Number 4: How much?—Annual TTI
Number 5: Who is responsible?—Priority owners

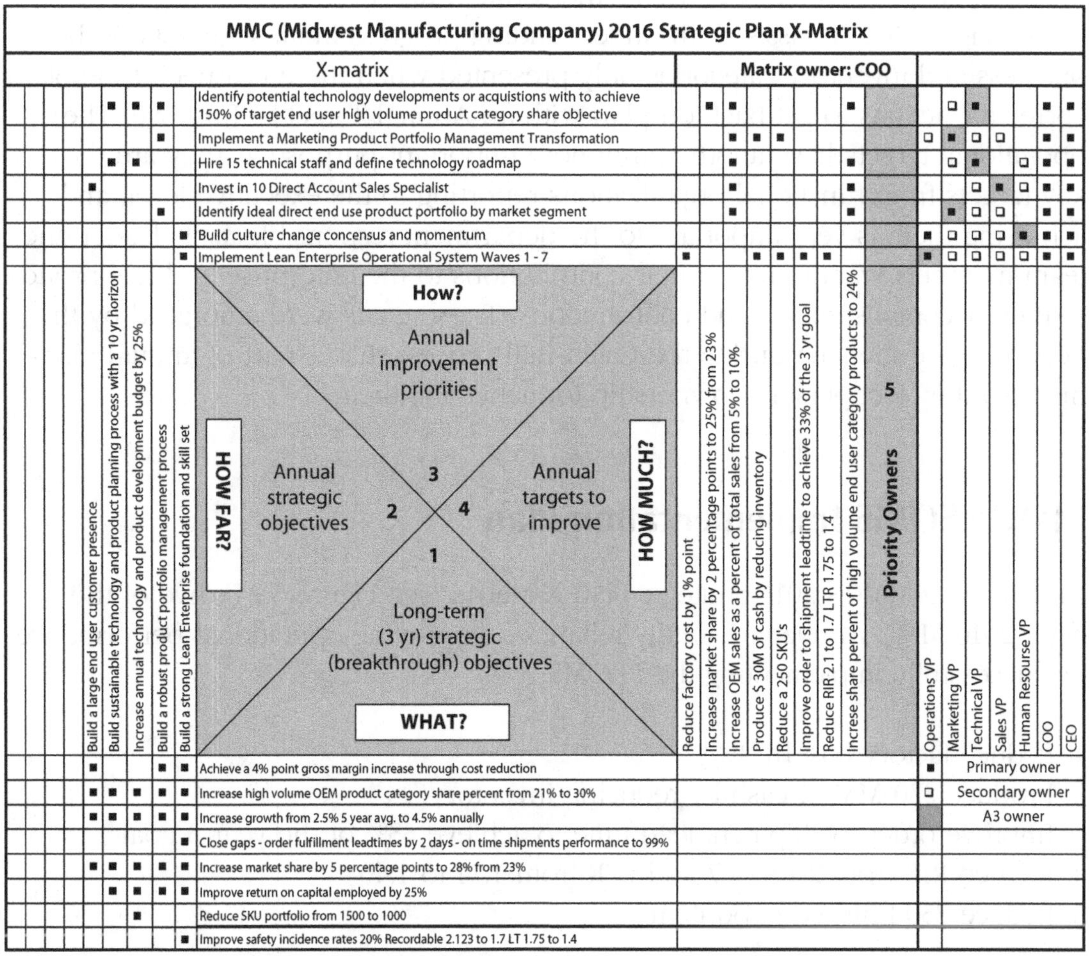

Figure 5.24 MMC 2016 Strategic Plan X-Matrix.

Annual strategic plan implementation dashboard																			
Each target you have previously entered will appear below automatically. For each target, enter your starting point and plot the progression needed to meet the target on time. Fill in the actually progression over time. Current cell formatting will show																			
Project status	Last status review date:			Green	On-time	Yellow	At risk	Red	Off-plan	Initiate		Continue							
Priority	Deliverables and measures	Starting point	Month	1	2	3	4	5	6	7	8	9	10	11	12				
1	Reduce RIR 2.1 to 1.7 LTR 1.75 to 1.4	2.1% R.I.R. 1.75% L.I.R.	Plan																
			Actual																
2	Reduce factory cost by 1% point	70.3%	Plan																
			Actual																
3	Increase share percent of high volume end user category products to 24%	21%	Plan																
			Actual																
4	Improve order to shipment leadtime to achieve 33% of the 3 yr 2 goal	5 days	Plan																
			Actual																
5	Increase market share by 2 percentage points to 25% from 23%	23%	Plan																
			Actual																
6	Increase OEM sales as a percent of total sales from 5% to 10%	5%	Plan																
			Actual																
7	Produce $ 30M of cash by reducing inventory	0$	Plan																
			Actual																
8	Reduce a 250 SKU's	450	Plan																
			Actual																

Figure 5.25   MMC 2016 strategic plan TTI Bowler.

Annual strategic plan "TTI" are brought to the Bowler (see Figure 5.25), which is reviewed by operations committee members monthly with the annual TTI owners, who are required to present corrective action plans to get any yellow or red targets back on track. In order to be actionable by MMC's organization, some of the strategic plan annual TTI are further broken down.

See Figure 5.26 to assist with understanding progress in all segments of the business and inventory categories only presented when there is a need to explain corrective actions to get back on plan with an Annual Target to Improve. The operational targets have another and more important purpose as they are the key targets from functional organizations reporting to the executive level, and achieving them is very important to the department responsible as its leader and team members want to make their contributions to the enterprise and feel proud of their accomplishments and contributions. These goals were embraced by the department leadership during the Catch-Ball process that is part of annual planning, so they feel personal ownership for achieving them.

MMC LEOMS Implementation Plan

In reviewing MMC's 2016 Strategic Plan X-Matrix (see Figure 5.24), five of the eight 2016 MMC Annual are totally reliant on improving operational performance by successfully implementing the LEOMS:

1. Reduce factory cost by 1%
2. Produce $30 MM of cash by reducing inventory
3. Improve order to shipment lead time to achieve 33% of the year 3 goal
4. Reduce RIR from 2.1 to 1.7 and LTR from 1.74 to 1.46
5. Reduce 250 finished good items

Midwest Manufacturing Company - MMC			
3-year strategic plan goals			**Assumptions**
	From	**To**	
Market share	23%	28%	Yr 1:1% Yr 2:2% Yr 3:2%
Change go-to-market model			**Market**
Distributors	20%	35%	Big regional distributors are gaining market share
Dealers	75%	45%	Dealers are losing share to regional distributors
Direct to end users	5%	20%	Large end users are rationalizing their supplier base
Product offering			
Rationalize SKU's	450	300	Eliminate extremely low volume products
Develop or acquire high volume category products	21%	30%	Must become more important to distribution and end users
Increase gross margins			**Gross margins**
Distributor sales	26%	30%	Wage inflation of 2.4% annually
Dealer sales	30%	34%	Material cost inflation 1.1%
End user sales	40%	44%	4% point increase to be achieved by cost reduction
Order entry to shipping lead time			
Distributor orders	5 days	3 days	Match best competitor
Dealer orders	5 days	5 days	Match best competitor
End user orders	4 days	2 days	Match best competitor
On time shipping performance			
Distributor orders	95%	99%	Be the standard
Dealer orders	94%	99%	Be the standard
End user orders	90%	98%	Be the standard
Inventory turnover			
Raw materials	3.5	5.0	Improve return on capital employed
In-process	8.0	18.0	Improve return on capital employed
Finished goods	4.0	8.0	Improve return on capital employed
Safety			
Recordable incidence rate	2.12	1.7	Reduce recordable incidence rate 20%
Lost time incidence rate	1.75	1.4	Reduce lost time incidence rate 20%

Figure 5.26   MMC 2016 strategic plan operational targets.

In addition, the LEOM's implementation is also supported by two annual strategic plan improvement priorities (see Figure 5.27):

1. Build culture change consensus and momentum
2. Implement LEOMS waves 1–7

In this chapter, (see Figure 5.28) the LEOMS's implementation model, "Wave 1: strategic alignment," has been discussed; the remaining waves 2–7 will be discussed in Chapters 6, 7, and 8.

MMC's LEOMS implementation performance dashboard (see Figure 5.29) reflects completion of wave 1, and it is time to get moving on planning and executing wave 2.

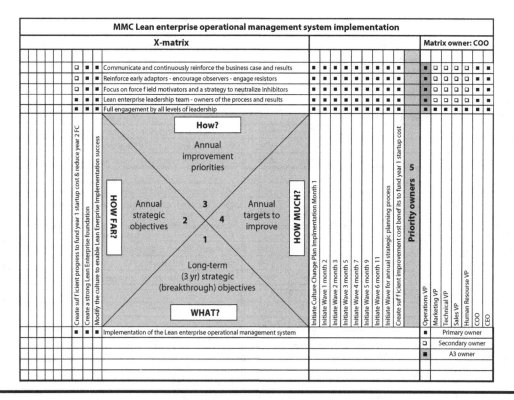

Figure 5.27 MMC LEOMS implementation.

Lean enterprise operational management system implementation model

Business unit operations implementation waves

Wave 1 Strategic alignment	Wave 2 VSM assessment and future state	Wave 3 Create the foundation	Wave 4 Lean work cell redesign	Wave 5 System deployment	Wave 6 Continuous improvement	Wave 7 Annual plan deployment
3 year Lean deployment plan Critical few metric targets: • Safety • Quality • Cost • Service • Inventory Lean transformation business case Lean transformation communications Launch culture change plan Launch Lean transformation LEOMS implementation Bowler project tracking	Lean introduction training VSM assessment • Material flow • Information flow • Disconnects • Performance measurements • Constraints • 8 wastes • Takt time • Spaghetti diagram • Mass balance VSM future state project portfolio LEOMS implementation A3 projects Yokoten Initiate plan for every part	5S Training 5S Application 5S Standards and check list 5S Scorecard 5 Min start of shift meetings Current and future state VSM implementation plan Current and future plant material flow map Facility 5 year flow plan Initiate suggestion system Daily audits PDCA Glass wall Yokoten	Work cell improvement Lean flow training simulation Improvement tools: • Work cell design • Standard work • Quality at the source • SMED • Andon • TPM • Error proofing • Day by hour chart • PDCA • Purchased materials Supermarkets and Kanban Yokoten	Value stream flow Level scheduling Pacemaker scheduling Kanban system Supermarkets Changeover wheel Material delivery routes Cross-training and certification Daily Kaizen Daily layered audits Yokoten	Executive led quarterly Lean reviews and plant "Go See" visits Bowler green-yellow-red charts Countermeasures and corrective actions Next quarter improvement plan Yokoten Repeat the process from wave 2 to wave 6 for system level improvements Leaders coach and live the Lean culture	Annual strategic plan deployment Hoshin Kanri planning • X-Matrix • True North • Catch ball • A3 • Bowler-Green-Yellow-Red • Periodic business reviews with PDCA discipline

1. Create goal alignment, define business drivers, critical few metric goals, create facility deployment and communication plans
2. Assess value streams and create future state visions for our facilities, value streams, processes, and performance
3. Lean system foundational elements which are implemented or in the process of being implemented
4. Optimize work cell processes, operator value added time and balance of work to create stability, flow, and improved performance
5. Schedule and synchronize production and material flow across the facility creating a complete Lean operational system
6. Quarterly Lean operational reviews to support continuous improvement and provide coaching to strengthen system knowledge
7. Integrate and align three year and annual strategic plan performance goals and and long-term programs into annual execution plans

Figure 5.28 LEOMS seven implementation waves.

Lean operational management system implementation performance dashboard																											

For each target, enter your starting point and shade in the number of months to complete each wave. Monthly plot progress in achieving the milestones for each month. If progress has not achieved the milestones a corrective action plan is required to be implemented to get back on plan.

Priority	Deliverables and measures	Starting point		1	2	3	4	5	6	7	8	9	10	11	12	13	14	15	16	17	19	20	21	22	23	24
1	Initiate culture change plan implementation month 1		Plan																								
			Actual																								
2	Initiate wave 1 month 2		Plan																								
			Actual																								
3	Initiate wave 2 month 3		Plan																								
			Actual																								
4	Initiate wave 3 month 5		Plan																								
			Actual																								
5	Initiate wave 4 month 8		Plan																								
			Actual																								
6	Initiate wave 5 month 11		Plan																								
			Actual																								
7	Initiate wave 6 month 12		Plan																								
			Actual																								
8	Initiate wave 7 for annual strategic planning		Plan																								
			Actual																								
9	Create sufficient improvement cost		Plan																								
			Actual																								

Figure 5.29 LEOMS's implementation dashboard.

LEOMS Implementation

Adopting the LEOMS

Becoming a Lean enterprise entails embracing new management thinking and a new operational system. It is a commitment to a complete transformation of how the organization thinks, plans, and acts in running the business. If this is just thought about as a set of practices that are bolted onto businesses, it is not worth doing. While some things might be slightly improved, it creates a condition of false hope and eventual disappointment, as improvement is not sustained, and eventually placement of blame on the LEOMS as something that does not work in "my business." Becoming a Lean enterprise is a forever journey toward perfection and should be thought about very seriously before committing everything, which is a necessary condition for success. This is particularly applicable to C-level business leadership as they may view Lean as something their organization should do but has little to do with them. Every C-level leader must embrace the required changes to culture and their thinking, and must see adoption of the LEOMS as transformational not only for their organization but also for themselves. Lean enterprise transformations are investments that will pay for themselves in the first 2 to 3 years and generate a significant financial returns in years following. Required investments for success are the following:

1. Training investments are normally accomplished through hiring outside Lean consultants to train and coach the organization; this can also be done by recruiting an experienced Lean operations executive and Lean coaches.
2. A dedicated Lean transformation executive is a requirement supported by Lean functional executive sponsors, who are implementation owners in their area of responsibility, keeping their organization accountable for learning, doing, and achieving results from Lean.

3. Full-time Lean champions/coaches, who drive the LEOMS deployment on a day-to-day basis, are a necessary part of every company's transformational investment.

4. Lean process owners, leaders of organizational units like plants for example, must be in place and taking active ownership for deployment of the LEOMS within their organizations.

5. Finally, organizations need to build their own internal Lean knowledge experts, often called Lean coaches. Coaches are normally full-time assignments for 2 or 3 years, learning from contracted outside consultants and expected to be knowledge and practice experts to sustain long-term Lean organizational progress and maturity. This is a great people development assignment and should be given to at company's best talent; this increases sustainable success odds and sends a clear message to the organization that being a "Lean leader and coach" is required for career advancement.

References

1. Kim S. Cameron and Robert E. Quinn, *Diagnosing and Changing Organizational Culture*, pp. 11–12, San Francisco, CA: Jossey-Bass, 2011.
2. Mike Rother, *Toyota Kata*, Burr Ridge: McGraw-Hill, 2010.
3. Steven J. Spear, *The High Velocity Edge*, Burr Ridge: McGraw Hill, 2009.
4. Satoshi Hino, *Inside the Mind of Toyota*, New York: Productivity Press, 2006.
5. Taiichi Ohno, *Toyota Production System*, New York: Productivity Press, 1988.
6. Jeffrey K. Liker and Michael Hoseus, *Toyota Culture*, Burr Ridge: McGraw-Hill, 2008.
7. John Shook, *Managing to Learn*, Lean Enterprise Institute.
8. Larry Bossidy, *Confronting Reality*, Random House, 2008.
9. Miller, Wroblewski & Villafuerte, *Creating a Kaizen Culture*, McGraw Hill, 2014.
10. Masao Nemoto, *Total Quality Control for Management*, New Jersey: Prentice Hall, 1987.

Chapter 6

Lean Enterprise Operational Management System Factory Operations Implementation

What Will I Learn?

How to implement and sustain the LEOMS, step by step.

The journey through this book started with a look at manufacturing history and its contributors who impacted operational system development, including how Taiichi Ohno built his innovation by understanding the manufacturing system's current strengths and limitations. He applied practical innovations to create a system to design, operate, and manage all types of supply chain models regardless of scale, technology, volume, or mix requirements. The Lean Enterprise Operational Management System (LEOMS) is a company's best investment to improve their odds of not just surviving but also thriving long term. This is a strong argument as to *why* every company should implement the LEOMS. Next, a detailed response is provided as to *how* the LEOMS delivers on its promise to continuously improve a company's value proposition execution and reduce cost. To be of great sustaining value, operational systems must continually enable and enhance an enterprise's strategic value proposition. The LEOMS accomplishes this with powerful and effective processes and tools starting with Hoshin Kanri, a proven process and tool set to ensure that both operational improvement and longer-term strategic capabilities are being improved year after year. Achievement of strategic and operational plan critical few metric targets is accomplished through direct linkage to frontline operational process improvement targets and projects focused on every day until they are achieved. Successful implementation of the LEOMS is neither free nor easy but is worthwhile, starting with examination of company culture to understand its compatibility with the LEOMS's people management culture requirements. Culture assessments identify critical cultural obstacles to success, those attributes that are not compatible with the LEOMS's kaizen culture. Culture change is a very big challenge because we all understand "Rule One" about culture; the big culture wins 99% of the time and crushes any "invaders," so the big culture must enable the LEOMS's success. With an effective culture change plan in place, the LEOMS implementation journey continues by applying the LEOMS's Waves 2 through 7, a structured process providing a logical step-by-step sequence for effective and comprehensive execution of LEOMS implementation. While the LEOMS is better than any other operational management system in existence, its value stream mapping (VSM) assessment tool is not the best choice for broader supply chain continuous improvement nor for transactional business function processes. These two important business process segments will be treated in Chapters 7 and 8 by applying well established assessment tools designed for broader supply chain and transactional business processes. These applications will include applying the LEOMS's continuous improvement thinking and its eight wastes to drive enterprise-wide continuous improvement. In Chapter 7, SCOR or the supply chain operational reference model[1] and its defined assessment tools and best practices along with supply chain mapping will be applied to create current and future states leading to identification of improvement projects to make the future state an operational reality. In Chapter 8, business process mapping will be applied to transactional processes to create a current state, future state, and current state with improvement projects to achieve future state performance improvements targets.

The Goal of LEOMS's Seven Waves

The LEOMS's (see Figure 6.1) implementation process is an eye opening and exciting time for those involved in learning to apply its tools and, most of all, seeing its results. It is engineers', team leaders', and group leaders' responsibility to be coaches and engage frontline production and maintenance team members in seeing opportunities and applying PDCA problem solving. It is important from day 1 to engage frontline team members so they feel ownership and a part of determining changes that will be made. At each step along the LEOMS's implementation journey, demonstrating the "new culture" is critical to sustaining continuous improvement and keeping frontline team members completely engaged. Yes, it is true that teams could go faster and get quicker results, at least in the short term, if engineers' and frontline team and group leaders' ideas were implemented, but the focus must be on long-term engagement of everyone from day 1 as this is the only way to build an army of improvement scientists. The organizational mentality should be like "rolling a big snowball down a hill"; being an upper Midwesterner, one has a lot of experience with snow and as kids playing in it was great fun. When snow and weather conditions were right, we would form and roll big snowballs down a hill and watch them pick up speed and smash into a tree and explode. Sometimes, when the

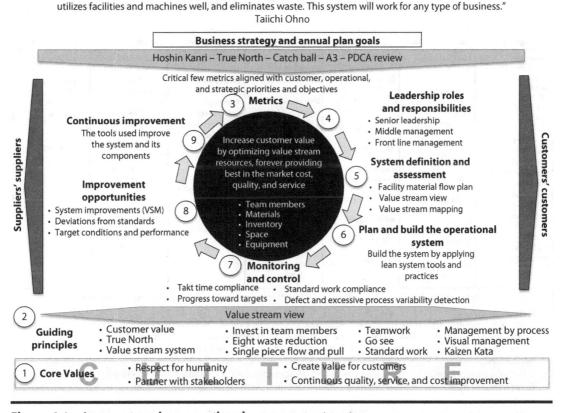

Lean enterprise operational management system
"A total management system is needed that develops human ability to its fullest capacity and fruitfulness, utilizes facilities and machines well, and eliminates waste. This system will work for any type of business."
Taiichi Ohno

Business strategy and annual plan goals

Hoshin Kanri – True North – Catch ball – A3 – PDCA review

Critical few metrics aligned with customer, operational, and strategic priorities and objectives

3 Metrics

Continuous improvement
The tools used improve the system and its components

9

Increase customer value by optimizing value stream resources, forever providing best in the market cost, quality, and service

Improvement opportunities
• System improvements (VSM)
• Deviations from standards
• Target conditions and performance

8

• Team members
• Materials
• Inventory
• Space
• Equipment

4 Leadership roles and responsibilities
• Senior leadership
• Middle management
• Front line management

5 System definition and assessment
• Facility material flow plan
• Value stream view
• Value stream mapping

6 Plan and build the operational system
Build the system by applying lean system tools and practices

7 Monitoring and control
• Takt time compliance
• Progress toward targets
• Standard work compliance
• Defect and excessive process variability detection

Suppliers' suppliers

Customers' customers

2 Value stream view

Guiding principles
• Customer value
• True North
• Value stream system

• Invest in team members
• Eight waste reduction
• Single piece flow and pull

• Teamwork
• Go see
• Standard work

• Management by process
• Visual management
• Kaizen Kata

1 Core Values
• Respect for humanity
• Partner with stakeholders

• Create value for customers
• Continuous quality, service, and cost improvement

CULTURE

Figure 6.1   Lean enterprise operational management system.

temperature was slightly above freezing, the snowball would pick up every flake of the snow, leaving bare ground in their path. This is the LEOMS's implementation goal; everyone needs to be part of its big snowball and no one should be left behind; that means managing the "weather" and the "snowball," so all the "snowflakes" are included as the big snowball increases speed going downhill.

LEOMS's Wave Implementation Model

This chapter will explain how to use the LEOMS's implementation model by applying its processes and tools as described in waves 2 through 7 (see Figure 6.2) to begin realizing real financial and strategic benefits. Just as a reminder, wave 1 was covered previously in Chapter 5. The implementation process must take into consideration specific company situations and circumstances. Company scale, number of plants, global plant locations, and planned investment in resources are factors in determining a reasonable timeframe to be able to complete a first pass through the LEOMS's Seven Waves. Since most transformations are not launched at the beginning of a calendar year, wave 1 initiates the implementation process and application of waves 2 through 6 would be applied

	Business unit operations implementation waves					
Wave 1 **Strategic** **alignment**	**Wave 2** **VSM assessment** **and future state**	**Wave 3** **Create the** **foundation**	**Wave 4** **Lean work cell** **redesign**	**Wave 5** **System** **deployment**	**Wave 6** **Continuous** **improvement**	**Wave 7** **Annual plan** **deployment**
3 year Lean deployment plan Critical few metric targets: • Safety • Quality • Cost • Service • Inventory Lean transformation business case Lean transformation communications Launch culture change plan Launch Lean transformation LEOMS implementation Bowler project tracking	Lean introduction training VSM assessment • Material flow • Information flow • Disconnects • Performance measurements • Constraints • 8 wastes • Takt time • Spaghetti diagram • Mass balance VSM future state project portfolio LEOMS implementation A3 projects Yokoten Initiate plan for every part	5S Training standard work 5S Standards and check list 5S Scorecard 5 Min start of shift meetings Current and future state VSM implementation plan Current and future plant material flow map Facility 5 year flow plan Initiate suggestion system Daily audits PDCA Glass wall Yokoten	Work cell improvement Lean flow training simulation Improvement tools: • Work cell design • Standard work • Quality at the source • SMED • Andon • TPM • Error proofing • Day by hour chart • PDCA • Purchased materials Supermarkets and Kanban Yokoten	Value stream flow Level scheduling Pacemaker scheduling Kanban system Supermarkets Changeover wheel Material delivery routes Cross-training and certification Daily Kaizen Daily layered audits Yokoten	Executive led quarterly lean reviews and plant "Go see" visits Bowler green-yellow-red charts Countermeasures and corrective actions Next quarter improvement plan Yokoten Repeat the process from wave 2 to wave 6 for system level improvements Leaders coach and live the Lean culture	Annual strategic plan deployment Hoshin Kanri planning • X-Matrix • True North • Catch ball • A3 • Bowler-Green-Yellow-Red • Periodic business reviews with PDCA discipline

1. Create goal alignment, define business drivers, critical few metric goals, create facility deployment and communication plans
2. Assess value streams and create future state visions for our facilities, value streams, processes, and performance
3. Lean system foundational elements which are implemented or in the process of being implemented
4. Optimize work cell processes, operator value added time and balance of work to create stability, flow, and improved performance
5. Schedule and synchronize production and material flow across the facility creating a complete Lean operational system
6. Quarterly Lean operational reviews to support continuous improvement and provide coaching to strengthen system knowledge
7. Integrate and align three year and annual strategic plan performance goals and long-term programs into annual execution plans

Figure 6.2   LEOMS implementation model.

sequentially after wave 1. Wave 7 is done during the organization's normal annual planning cycle.

Implementation Schedule

The LEOMS's implementation schedule example shown in Figure 6.3 is only an illustration as the speed and time to work through implementation depends on the factors previously discussed. This example would be appropriate for a company with a number of plants; of course, the amount of resources invested to make it happen would be another important factor relative to implementation speed. At a minimum, there needs to be a rhythm established that pushes the organization just a little beyond what it perceives as its limits. The days of "business as usual" are over, and the focus must be on confronting defects and barriers to achieve target state performance levels through rigorous daily continuous improvement.

	For each target, enter your starting point and shade in the number of months to complete each wave. Monthly plot progress in achieving the milestones for each month. If progress has not achieved the milestones a corrective action plan is required to be implemented to get back on plan.																											

Priority	Deliverables and measures	Starting point		1	2	3	4	5	6	7	8	9	10	11	12	13	14	15	16	17	19	20	21	22	23	24
1	Initiate culture change plan implementation month 1		Plan																								
			Actual																								
2	Initiate wave 1 month 2		Plan																								
			Actual																								
3	Initiate wave 2 month 3		Plan																								
			Actual																								
4	Initiate wave 3 month 5		Plan																								
			Actual																								
5	Initiate wave 4 month 8		Plan																								
			Actual																								
6	Initiate wave 5 month 11		Plan																								
			Actual																								
7	Initiate wave 6 month 12		Plan																								
			Actual																								
8	Initiate wave 7 month 12 annual strategic and operational plan		Plan																								
			Actual																								
9	Create sufficient improvement to cover year 1 startup cost		Plan																								
			Actual																								

Figure 6.3 LEOMS implementation dashboard.

LEOMS's Leadership

There are no success stories of radically improved results from transformational change not championed by a company's top leaders. Leaders do things they understand and believe because failing may cost them their jobs. This raises a question: how many CEOs understand the LEOMS well enough to bet their jobs and their companies on implementing it? The answer is very few. Leaders need to become familiar enough with the LEOMS to lead implementation because today's fast-moving global markets require an effective operational management system that includes enterprise-wide continuous improvement. Leaders who do not implement one risk their and their team members' jobs.

STANDARDIZE WORK AT ALL LEVELS

The introduction and early deployment of LEOMS within 3M manufacturing was supported by Rick Harris of Harris Lean Systems who along with the 3M manufacturing leadership conducted operational Lean reviews with all 60 facility managers in North America—6 to 10 facilities per session, which were held every 4 months. Harris devoted one of the sessions to teaching standardized work for managers.

Paul Husby

Daily frontline team member engagement with management is essential for maintaining operational discipline, continuous improvement, and employee development. The LEOMS, for example, is a people and leadership based process with layered audit control plans. Frontline team members and their leaders have a daily standardized work routine to audit certain areas. Plant managers spend their first hour or two each day reviewing frontline operations to audit, coach, and identify next areas of system improvement.

■ *Audit*: Leaders at all levels audit an area that others have audited to maintain consistency and ensure accountability because everyone knows that at some point their work is inspected. Audits are a preventive act, an early detection system for identifying small deviations that may cause quality, service, safety, or cost failures if not corrected.

■ *Coach*: Lean teaches practitioners to see with *new eyes* and seek perfection, skills that are fundamental to continuous improvement. Tools and methods obviously assist in this process, but a greatly increased sensitivity to waste and its sources will change managers' thinking. This skill needs to be coached and nurtured everyday by questioning, for example, why the hourly production target was not achieved, what countermeasures have been tried and what was learned, why something is done in the current way. Why, why, why, why? This drive to understand, identify root causes, and implement solutions is also an opportunity for leadership to learn and share their experience while building relationships with frontline team members. This does not include a habit of many executives of making suggestions of how to solve a problem. Instead, it is to question constructively what team members have discovered about their problem, what has been considered regarding possible root causes, or what countermeasures have been tried, because building an army of problem solvers will not be achieved if they are dependent on management to solve their problems.

■ *Identify next levels of improvement*: In addition to auditing and coaching, leaders must spend time observing operations not only to understand the details of a process but also to see how the entire system is functioning. This experience reinforces something most leaders know but do not do, which is that time dedicated to operational reviews and floor walks is critical to both process improvement and team leader, group leader, and frontline team member development. They must constantly remind themselves that they must be learners and problem-solving guides with patience to allow frontline team members to wrestle with problems, try, and fail repeatedly until they find the real cause and its solution.

Developing Team Members into an Army of Problem-Solving Scientists

Leaders as Teacher and Coaches

Lean leaders are teachers and coaches (see Figure 6.4) as the LEOMS only works effectively when the entire organization understands and executes their assigned roles in a manner that is continually reinforcing and improving their system. This is a critical factor in LEOMS success as it builds a stronger problem-solving army every day, resulting in consistent period over period operational performance improvement. Lean leadership coaching involves dedicating regularly scheduled time to "go see" by being on the frontline engaged with team members to see their problems and engage them:

- Ask questions about team member's job responsibilities and activities.
- What problems are you having?
- Why is this happening?
- What temporary or permanent countermeasures have been considered?
- Why is the implemented countermeasure the best choice?
- Show respect for team members' work.
- Verify that clear responsibility is assigned for every process and problem.

In his book *Toyota Production System*, Taiicho Ohno describes the power of Toyota's Production System management training as Ninjutsu, the art of the invisible (see Figure 6.5). Toyota Production System's magic effect is seen daily, as an army of problem solvers identify and solve problems, improving an operation's quality, cost, and service every day. This happens because of Toyota's leadership training and expectations regarding development of their people daily. The "magic" appears as team members follow their standard work, identify defects, apply scientific problem-solving methods, and share their learning with others who may have a similar application. Coauthor Jerome Hamilton confirms his experience at NUMMI Motors when

Rational:
1. Leaders are the most skilled at demonstrating and teaching
2. Leaders create learning organizations
3. Leaders need to create value and eliminate waste as it is the most important factors in determining their contribution and success

Responsibilities:
1. Organization/system design
2. Developing other leaders
3. Creating a "run to standard and target" culture
4. Continuous improvement of the organization and system

Figure 6.4   Leaders as teachers.

Ninjutsu

Ninjutsu is the art of invisibility. It means acquiring skills through training. Management must be done by Ninjutsu, the art of invisibility. As children we watched ninjutsu tricks in movies like the hero suddenly disappearing. As a management technique, however, it is something very rational. It means acquiring and applying management skills by training.
Toyota Production System –Taiichi Ohno, Productivity Press, 1988, pg. 69

We observe Ninjutsu results when reviewing the continued improving competitiveness of Toyota based on daily kaizen throughout their organization. It is the unseen magic that lies behind the Toyota paradox. Toyota revolutionized the motor vehicle industry in the past 40 years and continues to focus on daily kaizen to support their leadership position. The key to their success is training all levels of management to build and sustain the operational system including daily kaizen by all team members. While their training is not always observable on the shop floor, the engrained behavior of continuous improvement is observable.

Figure 6.5 Ninjutsu—management magic.

he was trained in the LEOMS tools and methods while being exposed to Toyota empowerment of frontline team members. As he states, it takes time for people to understand what empowerment means working within Toyota's operational management system. Their practice of salaried staff and management spending time working on the frontline is very valuable in order to get to know frontline team members and the challenges they face in order to be effectve in coaching them.

Train the Trainers

The LEOMS's implementation success depends on effective Lean coach development; all leaders are coaches and need to be trained to ensure short- and long-term success as it depends on their competence, leadership skills, training, and coaching ability. All leaders are expected to model behaviors in their daily

---

NUMMI INVESTING IN PEOPLE

One of the things I learned at NUMMI was the importance of investing in people. When companies embark on Lean journeys, the word empowerment is often used. I think some people misunderstand empowerment. I've heard many floor operators say, aren't we suppose to be empowered? The answer should be "Yes, you're empowered to help deliver the business results with in our system" Therefore, people must be well trained and engaged enough to understand what the key goals and objectives are. At NUMMI it was mandatory for me to take courses like practical problem solving, PDCA, to learn how to create standard work and apply Hoshin Kanri. I also had to work the production floor an entire day in each department to gain a better understanding of what the operators were experiencing. This allows for better decision making.

Jerome Hamilton
Investing in People at NUMMI

1. Engages people at all organizational levels

2. Teaches team members to focus on the work, material flow, and value stream to see the waste

3. Gives team members deep technical and process knowledge

4. Pushes responsibility for value stream management and improvement to the lowest possible line management level

5. Introduces metrics to encourage horizontal thinking

6. Creates frequent problem solving loops between themselves and their superior and themselves and their subordinates

7. Accomplishes these six leadership responsibilities through application of policy deployment, A3 analysis, standardized work with standardized management and kaizen

Figure 6.6 LEOMS leader people leadership expectations.

interactions with team members consistent with the LEOMS's people development expectations listed in Figure 6.6. This document should be very visible and discussed frequently with all team members both to reinforce leaders' commitment to the LEOMS's implementation process and to illicit constructive feedback during crew meetings and frontline team member conversations. Great leadership starts with leader authenticity, which is demonstrated by their willingness to hear and take corrective action on areas that can be improved; doing this is a powerful way to gain team member respect and support. This message, lived out each day in interactions with team members, reinforces the recognition that leaders are open to improving and are legitimately concerned with each team member's development. These expectations should also be a major part of leader performance evaluations and promotion criteria. Leadership evaluations from team members under a leader's responsibility should count as the most significant factor in rating leaders' annual performance. The LEOMS's people management focus on engagement and investment in frontline team members is a leader's most important contribution to LEOMS sustainable implementation success.

Team Members' Organizational Role and Responsibilities

Frequently, Lean enterprises present their organization charts with frontline team members at the top as a reminder that unless leaders coach and support their team members and lead continuous improvement every day, sustainable success will not be achieved. It is also important to review and absorb frontline team members' role within the LEOMS (Figure 6.7). The Ninjutsu will only appear consistently when frontline team members own and practice their role every day, so they begin to believe in the system, own it, and practice their part daily. Coaching, encouraging, and congratulating team members as they develop their problem solving skills and adopt standard work and other system elements are vital to building momentum and getting the snowball to the top of the hill, so it can start rolling downhill from its own momentum. See Figure 6.8 roles

1. Set their own goals
2. Achieve and maintain perfect attendance
3. Follow standard work:
    1. Do the job right
    2. Stay within takt time
    3. Work in a safe manner
    4. Follow the same sequence
4. Follow instructions
5. Know all the aspects of the job:
    1. Tally sheets
    2. Terminology
    3. Repair ticket
    4. Specifications/C.V.I.S.
    5. Options
    6. Equipment functions
6. Self motivated
7. Responsible for your work area
8. Safety
9. Positive attitude
10. Ability to accept and support change
11. Maintain high standards
12. Communicate with: (be tactful)
    1. Process team members
    2. Team and group leader
    3. Manufacturing management
13. Be involved to:
    1. Eliminate muda
    2. Continuous improvement – Kaizen
    3. Submit suggestions
14. Participate in problem solving
15. Be on time and ready to go to work
    1. At your station
    2. Have all job material
    3. Have gloves, tools, pens, stamps, etc.
16. Follow the code of conduct
17. Attend quality meetings
18. Attend all team meetings
19. Train and assist in training others
20. Willingly rotate based on quality and training status
21. Be a team player
22. Verify your time card before signing it.
23. Follow vacation procedure.
24. Be able to work with minimal supervision
25. Take pride in your work and in the product
26. Be willing to work overtime:
    1. To support repair
    2. To support production
    3. To support training
    4. To support quality problems
    5. To attend team group meetings

Figure 6.7 Team member role and responsibilities.

Roles and responsibilities by wave – realizing ninjutsu

Train – Apply – Practice – Verify – Coach

	Strategy, alignment and launch	Resource, train, and assess	Create foundation	Work cell redesign	System deployment	Sustain continuous improvement	Annual plan deployment
Team members			✓ Lean orientation ✓ Lean 5S training ✓ 5S participants	✓ Lean training ✓ Project participants ✓ Kaizen participants ✓ Initiate improvement ideas	✓ Lean training ✓ Project participants ✓ Kaizen participants ✓ Initiate improvement ideas	✓ Recognize waste ✓ Project participants ✓ Initiate improvement ideas ✓ Kaizen practitioner	✓ Recognize waste ✓ Kaizen practitioner ✓ Support the team
Team and group leaders		✓ Lean orientation ✓ Lean system training ✓ VSM assessment participation	✓ Lean training ✓ 5S participants	✓ Lean training ✓ Project participants ✓ Kaizen participants	✓ Lean training ✓ Project participant ✓ Kaizen participant ✓ "Evangelist"	✓ Lean training ✓ Project participant ✓ Kaizen participant ✓ Evangelist"	✓ Project leader ✓ Kaizen leader ✓ Evangelist" ✓ Yokoten facilitator
Lean coaches	✓ Select and train lean coaches	✓ Coach training ✓ VSM assessment leader ✓ Be the evangelist ✓ Facilitate Yokoten	✓ Coach training ✓ 5S launch ✓ Knowledge transfer ✓ Facilitate Yokoten	✓ Coach training ✓ Project leader ✓ Knowledge transfer ✓ Facilitate Yokoten	✓ Coach training ✓ Kaizen and project leader ✓ Knowledge transfer ✓ Facilitate Yokoten	✓ Coach training ✓ Kaizen and project leader ✓ "Evangelist" ✓ Knowledge transfer ✓ Facilitate Yokoten	✓ Train and develop project teams and employees ✓ Lean coaching ✓ Facilitate Yokoten
Leadership	✓ Align business strategy and Lean ✓ Communicate case for change ✓ Implementation timeline ✓ Culture change plan ✓ Commit resources	✓ Executive training ✓ Apply Hoshi Kanri ✓ Prioritize resources ✓ Critical few metrics ✓ Bowler progress tracking	✓ Executive training ✓ Bowler progress tracking ✓ PDCA ✓ Drive Yokoten across facility operations ✓ Shop floor visibility	✓ Executive training ✓ Bowler progress tracking ✓ PDCA ✓ Drive Yokoten across facility operations ✓ Shop floor visibility ✓ Communicate successes	✓ Executive training ✓ Bowler progress tracking ✓ PDCA ✓ Drive Yokoten across facility operations ✓ Shop floor audits and coaching ✓ Communicate successes	✓ Quarterly executive lean review ✓ Bowler progress tracking ✓ Drive collaboration in the business ✓ Shop floor visibility ✓ Communicate successes	✓ Integrate Hoshin-Kanri into annual planning process ✓ Host coaches quarterly lesson's learned ✓ Company wide communication ✓ Shop floor visibility
Lean training partners	✓ Train ✓ Lead first application ✓ Assess and coach leaders and teams	✓ Train ✓ Lead first application ✓ Assess and coach leaders and teams	✓ Train ✓ Lead first application ✓ Coach and assess leaders and teams	✓ Train ✓ Lead first application ✓ Assess and coach leaders and teams	✓ Train ✓ Lead first application ✓ Assess and coach leaders and teams	✓ Train ✓ Lead first application ✓ Assess and coach leaders and teams	✓ Train ✓ Lead first application ✓ Assess and coach leaders and teams

Figure 6.8 Roles and responsibilities by wave.

and responsibilities by wave, for all levels of an organization implementing the LEOMS. It is not an unreasonable idea during the first year of implementation to provide every team leader, coach, and front-line team member with a 3 × 5 notecard of key points related to their role and responsibilities and to encourage them to look at it every morning and carry it in their pocket as a reminder to "do it." There need to be intense external reminders and motivators until desired leadership behaviors become "Management Kata," routines that have been practiced repeatedly until they are "automatic," the norm for all leaders.

Training Method

The LEOMS's training method (Figure 6.9) is a simple and common-sense approach to providing training that "sticks." We have all heard the saying, "we need to hear something at least three times before it sticks with us" and that is the first training goal. The second issue is "the second law of thermodynamics"; all systems left without sources to maintain them will deteriorate into disorder. This means the training process is never done as it must be reinforced by practices such as standard work that are kept in order by the hierarchy of relevant checks, carried out by team members and leadership, verifying that systems are being operated and maintained as they were designed and specified.

Lean Coach Development Assessment

A high-potential top performing HR professional should administer Lean coach development assessments to provide coaches with feedback to assist them in their development. This assessment is to be completed at every organizational leadership level, as inconsistencies in modeling the culture will inhibit building a strong

Roles and responsibilities:

Lean coach and practitioners to conduct regular training/knowledge transfer at plant sites and identify tasks to be completed each month.

Train – Apply – Practice – Verify – Review

Lean team leaders will maintain a list of planned actions for the month and maintain over-all schedule/activities in their area of responsibility.

Lean coaches will share knowledge from all plants to accelerate learning and internalize Lean knowledge and implementation skills.

Lean coach at each facility is responsible for leading the implementation project including coordinating and training project teams. This coach will work with additional site resources as required to achieve the implementation schedule.

Plant management at each site will be the owner of the Lean Operational Management System transformation implementation schedule.

Figure 6.9   LEOMS's training method.

ACTIVITIES TO GENERATE ENTHUSIASM AND CONVINCE SCEPTICS

Two great approaches to helping people feel enthusiastic or overcome their skepticism by seeing the waste are calculating Machine OEE and doing "random" work sampling to identify the value added and non-value added percentages of operators' scheduled time on the shop floor.

Paul Husby

---

army of continuous improvement scientists. During the first couple of years of implementation (see Figure 6.10), assessments should be done twice a year. Once it is established that 80% of leaders are living the desired behaviors, development assessments should be performed annually and be included in annual performance reviews with clear improvement expectations set for the following year.

	360° Lean culture coach development assessment	Rarely	Sometimes	Consistently	Coach feedback guidelines
	Sensei - Coach other Coaches 156 - 195 Points - Continue to work on moving 3's to 5's				1) Keep suggestions positive by thinking about how you would respond if they were given to you
	Coach - 117 - 155 Points - Continue development to moving 1's to 3's and 3's to 5's				2) Provide specific examples as people can understand if it is a real life example
	Apprentice < 117 - Continue learning and development moving 1's to 3's	1	3	5	3) Maintain confidentiality by keeping your inputs to yourself as gossiping damages teamwork
	Respect for People - Leading by Example				**Provide concrete examples and constructive suggestions for improvement**
1	Consistently treats people with dignity and respect				
2	Encourages team members to contribute their ideas in team meetings				
3	Admits mistakes and encourages team members to admit mistakes				
4	Facilitates participation of team members in decisions that affect their work				
5	Listens attentively to team members' issues				
6	Focuses team members on finding root causes and not someone to blame				
7	Reaches conclusions, only after discovering the facts				
8	Gives recognition to the team's contributions to the larger organization				
9	Provides constructive performance feedback to team members				
10	Respects team members' time (i.e. starts and completes meetings on schedule)				
11	Regularly recognizes team members' contributions to the group's success				
12	Consistently rewards team members for their job performance				
	Continuous Learning and Development				
13	Motivates team members to continously improve their knowledge and work skills				
14	Makes development plans for team members, insuring they receive appropriate job training				
15	Teaches all team members coaching skills and leaders how to train and coach their teams				
16	Coaches team members and groups on job performance, consistently and constructively				
17	Encourages team members to look for and expose problems so they can be worked on and solved				
18	Takes a "System" perspective when looking for and making improvements				
19	Coaches and encourages a team approach to problem solving				
20	Encourages team members to share their ideas and knowledge with others				
21	Makes themselves available to assist teams from other areas in problem solving				
22	Supports team members when they need assistance				
23	Seeks out already proven solutions and promotes them with teams and team members				
24	Facilitates and encourages continuous change as good for the customer and job security				
	Lean Process and Methods Execution				
25	Engages team members in establishing work processes and standards				
26	Regularly validates adherance to work processes and standards				
27	Responds rapidly when problems are escalated = to her/his level				
28	Ensures disciplined applications of problem solving methods when problems arise - No Just do it's!				
29	Encourages team members to stop the process when quality problems arise				
30	Requires teams to deploy visual controls in all processes when appropriate				
31	Is disciplined to take action on team members' suggestions to improve the processes				
32	Makes decisions on a timely basis				
33	Is a good example by making decisions on data and facts not subjective information				
34	Follows up to ensure control plans are in place so solved problems don't reoccur				
35	Holds teams, leaders, and members accountable for executing their responsibilities				
36	Focuses on external and internal customers' needs and requirements				
37	Ensures team members receive needed information on a timely basis to be able to do their job				
38	Communicates regularly the business priorities and how team members can support them				
39	Creates an open communication environment so people feel free to express themselves				
Total					Team Member:
Notes:					Area/Dept: Date (MM/DD/YYYY):

Figure 6.10 Lean coach development assessment.

Wave 2

VSM Assessment and Future State

Wave 1 was dealt with in Chapter 5 providing MMC's Annual Supply Chain Operations Targets to improve, the critical few LEOMS implementation measures of success (see Figure 6.11) and the process is ready to move forward with wave 2.

Wave 2 LEOMS Implementation Dashboard

Wave 2 (see Figure 6.12) kicks off the hands-on parts of LEOMS implementation that continues through wave 6. During implementation of waves 2 through 6, great improvement ideas will arise that cannot be implemented because they are dependent on implementation of the future state plant layout and/or value stream

Each target you have previously entered will appear below automatically. For each target, enter your starting point and plot the progression needed to meet the target on time. Fill in the actual progression over time. Current cell formatting will show																		
Project status	Last status review date:			Green	On-time	Yellow	At risk	Red	Off-plan	Initiate		Continue						
Priority	Deliverables and Measures	Starting Point	Month	J	F	M	A	M	J	J	A	S	O	N	D			
1	Reduce RIR 2.1 to 1.7 LITR 1.75 to 1.63	2.1% R.I.R. 1.75% L.T.I.R.	Plan		On-plan													
			Actual	2.11 1.77	2.09 1.73													
2	Reduce factory cost by 1% point	68.0%	Plan		At risk													
			Actual	68	68.2													
4	Improve order release to shipment lead time by 1 day	5 Days	Plan		At risk													
			Actual	5	5.2													
7	Produce $ 30M of cash by reducing inventory	$0	Plan		On-plan													
			Actual	$0	$0													

Figure 6.11   Annual supply chain operations targets to improve.

For each target, enter your starting point and shade in the number of months to complete each wave. Monthly plot progress in achieving the milestones for each month. If progress has not achieved the milestones a corrective action plan is required to be implemented to get back on plan.																										
Priority	Deliverables and measures	Starting point		1	2	3	4	5	6	7	8	9	10	11	12	13	14	15	16	17	19	20	21	22	23	24
1	Initiate culture change plan implementation month 1	Plan																								
		Actual																								
2	Initiate wave 1 month 2	Plan																								
		Actual																								
3	Initiate wave 2 month 3	Plan																								
		Actual																								
4	Initiate wave 3 month 5	Plan																								
		Actual																								
5	Initiate wave 4 month 8	Plan																								
		Actual																								
6	Initiate wave 5 month 11	Plan																								
		Actual																								
7	Initiate wave 6 month 12	Plan																								
		Actual																								
8	Initiate wave 7 for annual strategic planning process	Plan																								
		Actual																								
9	Create sufficient improvement cost to fund year 1 startup cost	Plan																								
		Actual																								

Figure 6.12   LEOMS implementation dashboard.

improvements before they are feasible. These ideas should be documented so that they can be implemented during a future wave implementation cycle. Wave 2 focuses simultaneously on two major deliverables, plant material flow and value stream current states, and future states and current states with improvements initiating an annual cycle through waves 2 through 7.

Plant Material Flow Plan Target State

Improvement of value streams is the LEOMS's focus, but many plants have multiple value streams so it is not enough to only deal with value streams, as this may not optimize overall plant material flow. To avoid this issue, plant material flow planning should be undertaken to document the overall plant current state flow and a long-term plant material flow target state or future state (see Figure 6.13) that will optimize plant material flow. It is very common for Lean plants to create available space as over time they have changes in processes, materials, and finished products. When these changes occur, it can be a great opportunity to make facility changes to improve plant flow and efficiency, so it is prudent to have a target state plan to improve overall plant material flow.

To start, create the "current state" material flow plan by documenting flow using a "spaghetti diagram." It is a very simple tool to document material flow from the receiving dock, through storage, into production, out to finished goods storage, and finally to the shipping dock. Instructions for creating the diagram can be easily found by searching the Internet for the term "spaghetti diagram." Drawing the current state by hand on a flip chart or brown paper is the best way to construct a current state and brainstorm potential improvements. There is a multitude of computer applications also, but at least during initial documenting and brainstorming

VSM assessment and future state deliverables:
- Plant material flow spaghetti chart
- Current state value stream maps
- Future state value stream maps
- Current state improvement projects
- Plant material future state plan
- Plant material flow improvement projects
- Inventory plan – what?, how much?, where will it be stored?
- Headcount ideal state – including the roles for each individual
- Three year cost, productivity, delivery, safety, and inventory performance targets
- Portfolio of improvement opportunities documented on A3 documents
- Takt time
- PFEP

Information required
- Volume forecast by product – high (the dream); low (worst case); most likely (highest probability of happening)
- Product supply plan – products and volumes
- 3 year target performance of the key business and operational metrics
- Capacity and supply chain strategy – define based on demand profiles, volume seasonality, service strategy
- Plant facility drawings

Figure 6.13   Wave 2 VSM assessment and future state.

phases, it is best to brown paper, flip charts, and Post-It notes. Starting at receiving, trace material flows of purchased material categories to their storage locations, into production and in-process materials flow through production into finished good storage and to shipping. There are two purposes for spaghetti diagrams: first, they are a valuable visual tool for optimization of current material flow through a plant, and second, they are useful for creating "future state" and "target state" plant material flow plans. These are very valuable as plants go through technology and product changes over time, so having "a target state plan" will support making facility changes that enhance plant material flow. Every plant implementing the LEOMS will have opportunities to optimize their plant material flow after a couple of years of implementation. Material flow plan deliverables are a 3-year plant material flow target state, and current state, future state, and current state with improvements. Updated material flow plans should accompany each cycle through the LEOMS's waves 2 through 7 as new requirements and improvement ideas are generated.

Defining Value Streams

Defining value streams is accomplished using a product and process matrix (see Figure 6.14). Process and product matrices are constructed by documenting products or product families that share the same equipment. In Figure 6.14, products or product families are listed in the first spreadsheet column, and processes are listed sequentially at the top of sheet columns starting from the first process column on

Sort product families by routing similarity															
1. Look for patterns in the data															
2. Don't assume routings are fixed - alternatives may exist allowing products to be maybe a better overall option															
3. Take advantage of experienced process and process development engineers to assist in identifying the best routing option															
Operations / Part number	Saw	Engine Lathe	Lathe 1	Lathe 2	Lathe 3	Axis Mill 1	Axis Mill 2	Axis Mill 3	Hone	Super Hone	Studer Grinder	O.D. Grinder	I.D. Grinder	Balance	Speed Test
571747-1	1	2			3		4		5		6			7	8
204829-13	1	2		3			4		5		6			7	8
201798	1	2			3		4		5		6			7	8
202909	1	2			3		4		5		6			7	8
205424	1	2			3		4			5		6	6		8
2205639-1		1			2			3					4	5	
2205639-2		1			2			3					4	5	
201799		1			2			3					4	5	
202892			1			2			3					4	5
203595-1			1			2			3					4	5
203595-2			1			2			3					4	5
203093-2			1			2			3					4	5
2205639-3			1			2			3					4	5
201799	1	2		3			4								
203093-1	1	2		3			4								
202094-2	1	2		3			4								
202094-3	1	2		3			4								
203595-3	1	2		3			4								
Observations:															
Up to 30% process content difference can often be accommodated in the same value stream															
Sometimes significant differences can be accommodated by shuttling the different product through an off-line cell attached to the main flow line															

Figure 6.14 Product and process matrix.

the left. For each product, write sequential numbers starting with 1 if a product passes through the process shown at the top of the column. Continue this process until all processes are identified for all products. As noted in Figure 6.14, VSM's general rule is that products must not have more than 30% different processes to be included in the same value stream. Secondly, a suggestion in Figure 6.14 is when a significant number of processes are shared by products or product families having unique process stations, they may be accommodated by laying them out near a shared value stream process in a manner best supporting line flow and minimizing handling waste. It is advisable to engage process engineers, maintenance, and frontline managers in this process, as they typically are most familiar with how equipment is actually used and the products that flow through each process.

VSM—Assessing the System and Planning Improvement

Note: If you are already familiar with VSM, skip to the next section (Initiate PFEP).
VSM is the Lean system's assessment and planning tool. The best resource for learning about this specialized type of process mapping is *Learning to See*[2] by Mike Rother and John Shook (published by Lean Enterprise Institute). VSM includes a standard set of icons and instructions for completing value stream maps. Microsoft Visio supports an application called e-VSM that makes updating and maintaining value stream maps easy and less time consuming than redrawing them by hand. It is advisable to draw them by hand initially as it contributes to better implementation team VSM understanding and skill development. The VSM process (see Figure 6.15) starts by documenting the current state followed by creating a future state and finally defining a current state with improvements, noting the projects to be done to reach the future state. One- and three-year strategic goals for supply chain operations provide definitions of "Targets" state performance expectations. It is often advisable to develop 1- and 3-year future states as some beneficial changes may require capital investment that has not been included in the current annual budget. Continuously repeating the VSM process forever through improvement cycles focuses on the LEOMS's ultimate goal: perfection—zero waste.

1. Completing a current state map starts by documenting material and information flows based on actual frontline observation. Before actually documenting a current state VSM, teams must be trained in the LEOMS's VSM basics. Observers walk upstream, starting at shipping and noting key data about inventory, storage locations, and replenishment processes, and then continuing through each process operation and material handling process, and ending up in raw material storage and the receiving dock. Participants should note product flow through all process steps with key data such as the number of operators, output per day, cycle times, scrap, machine up times, and shifts of operation. Another excellent metric for value streams with automated process equipment that contributes to a significant portion

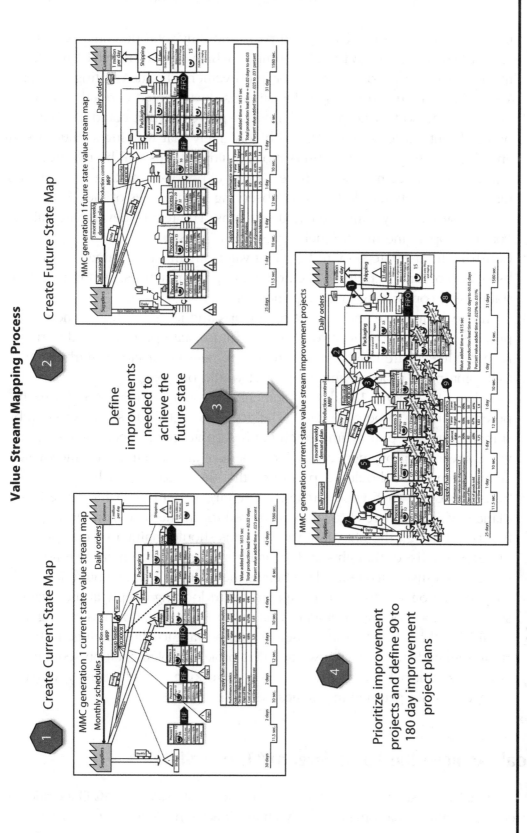

Figure 6.15 VSM assessment and future state process.

of the cost is the overall equipment effectiveness (OEE). It is a two-step calculation. Step 1 is calculated as follows: ((total monthly scheduled hours)/(process cycle time)) = monthly total potential number of parts. Step 2 is as follows: ((total monthly actual good part)/(monthly total potential parts)) = % OEE. Frequently, OEE is a low number, which should be celebrated as it means there is great opportunity for improvement. Teams creating current VSMs should focus on getting value streams documented correctly and with enough detail so that no potential areas or processes that could be sources of significant waste are missed. It is highly likely that walking a value stream and documenting data and observations multiple times will be required, particularly if a team does not have a lot of experience making value stream maps. Participants should focus on observing occurrences of the LEOMS's eight wastes as they walk their value streams. The teams' developmental goal is to open and mature their "eyes to see" waste.

2. Construct future state(s) by redesigning value streams to eliminate waste using appropriate Lean tools, methodologies, and practices, creating a future state vision conforming to Lean principles.

3. Compare future state and current state maps and identify improvement projects that will move toward achieving future state processes and performance. Projects are noted on a current state map by placing a starburst at the point in the process where a project would be done.

4. Develop a 90- and/or 180-day implementation plan including A3's for each potential project to be completed. The goal is to get value streams to exhibit smooth flow and pull by implementing high potential projects to realize value stream improvement and achievement of annual targets to improve. Teams doing VSM should keep in mind that their goal is to achieve the "target state" for each value stream. These targets define ideal state performance, which is a stretch goal to be strived for in the first year. During the first pass through the LEOMS's waves, teams should maintain a discipline of keeping a list of thoughts and ideas for system improvement that require some prerequisites to be in place before they can be implemented. It is easy to get bogged down, as once teams start looking deeply at their value streams, a flood of improvement ideas will be somewhat overwhelming. So organizing them in a logical sequence is important to allow change to be started immediately with knowledge that project implementation sequence is well thought out and will not result in reworking areas because a valuable change was not planned in a logical sequence.

Goal: Achieve the Value Stream "Target State"

MMC's initial value stream target state performance is shown in MMC Generation 1 current state value stream map's supply chain operations performance metrics (see Figure 6.16).

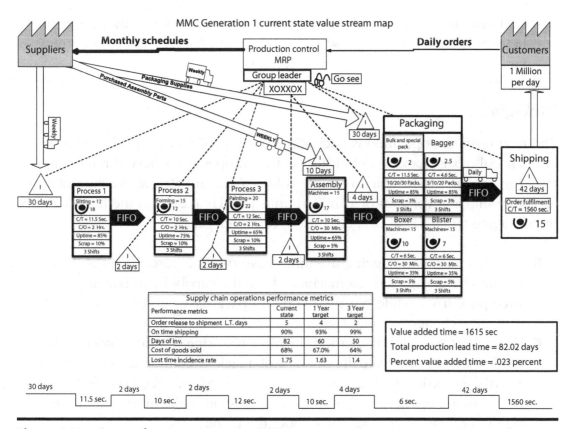

Figure 6.16 Generation 1 current state VSM.

Step 1. Value Stream Mapping Example: MMC

MMC has adopted the LEOMS as their path to sustainable success, and after thorough VSM training and practice, VSM Generation 1 current state was developed. Creating a value stream map always starts at shipping, the operation closest to their customers.

Shipping

At MMC, finished goods warehousing and shipping operations are in a separate building on the same site as the manufacturing plant operations:

■ Finished product from packaging is transferred each day to shipping where complete customer orders are assembled and shipped to customers.

■ Shipping has 42 days of finished goods inventory, and the average time to complete all value-added activities to fill an order is 26 minutes or 1560 seconds from the time an order filler picks up the order through the order being loaded on a delivery truck (see Figure 6.16). Normally, order entry through shipment lead time would be the desired metric, but MMC had never measured their order entry to shipment cycle time. It will be necessary at some point to measure the total process from order entry to shipment lead time and value-added time; this will be resolved when they apply the LEOMS's transactional improvement process in Chapter 8, as their current focus is on improving shop floor operations.

Packaging

The next upstream operation is packaging where products are packed in four formats: bulk, bags, boxes, and blister packs. All operations have throughput rates high enough to be able to handle volumes that are 50% above current average daily shipping volumes of 1 million units per day based on the historical mix of packaging types (bulk, bagged, boxed, or blister pack).

■ The bulk pack and bagging machines have runtime percentages at 85% and no changeover time as machines have the capability to bag any of three pack sizes, while boxing and blister pack have run times of 35% and 30-minute changeover times.

■ The current "scheduling method" involves checking frontline order status and inventory positions daily because of a lack of confidence in inventory records and order status reports. As a result, level material flow does not exist, causing operator and machine idle time, in addition to creating very high in process inventories throughout the plant.

Assembly

Assembly operations feed finished unpackaged product to packaging.

■ There are 15 semiautomatic assembly stations capable of changing over to produce any item in the product line in 30 minutes.

■ Current staffing level of 17 team members per shift and machine capacity on three shifts is sufficient to meet the average daily demand.

■ Total flow time through the plant is extremely long, resulting in production control spending significant time each day expediting and de-expediting orders to improve on-time shipments.

■ Assembly is scheduled each day based on a review of orders to be shipped and inventory in packaging and shipping. Day shift assembly and shipping team leaders provide production control with daily schedule recommendations after a visual review of their areas so orders for products in backorder can shipped.

■ Currently, the average assembled product inventory between assembly and packaging is 4 days. Inventory piles up ahead of assembly as scheduling of upstream operations are reprioritized daily by the MRP. MMC's forecast-driven push scheduling system is not well synchronized with daily customer demand and responsible for much of the inventory pile. Inventory between painting and assembly is 2 days.

■ Assembly is, in fact, not scheduled as it processes available material from forming and painting on a first-in first-out (FIFO) flow, along with production orders that are being expedited. This push process flow system makes it more difficult for MMC to respond to daily demand variability resulting in a lot of in-process inventory, daily chaos caused by constant pushing back and pulling forward shop work orders to meet customer demand each day.

Slitting, Forming, and Painting

Slitting, forming, and painting are scheduled daily by production control, and they suffer from similar issues as assembly including having 2 days of inventory between them, underutilized team members, low uptimes, and high scrap rates. The plant operates daily in a chaos and crisis mode responding to their latest fire. The two main slitting operation purchased material suppliers provide monthly shipments resulting in 30 days of inventory.

Current State Value-Added Time and Performance

MMC's total product flow time passing through the plant is 82 days, and the actual time of value-added processing is 26.9 minutes or 1615 seconds. In other words, plant product flow time is 118,053 minutes, and only 26.9 minutes or 0.00023% of the time is value added—118,080 minutes of waste! The Current State Supply Chain Performance Scorecard shows current state and one-year targets for the critical few supply chain metrics.

Step 2: Creating the Future State

Creating a future state map requires sufficient knowledge and application experience with LEOMS tools and practices to assess current state value stream processes, envision a future state, select appropriate practices and processes and determine the best application sequence to create a challenging but achievable future state. Generation 1 future state value stream defines the LEOMS's practices and tools to be applied to improve MMC's current state value stream. Appropriate LEOMS practices are specified at each operation to create a system of continuous end-to-end flow and pull based on customer orders. In their book, *Learning to See*,[2] authors Mike Rother and John Shook teach VSM to create value and eliminate waste, providing eight valuable questions to guide future state development (see Figure 6.17).

1. *What is Takt time?* Takt time establishes a value stream's operational rhythm. It is determined by dividing pacemaker operation's average daily customer demand by its daily available time. In MMC's case, assembly is the logical process to use for calculation of Takt time because that is where a product's final configuration is assembled. MMC's average demand is 50,000 per day and operations run three shifts having 75,600 seconds available each day;

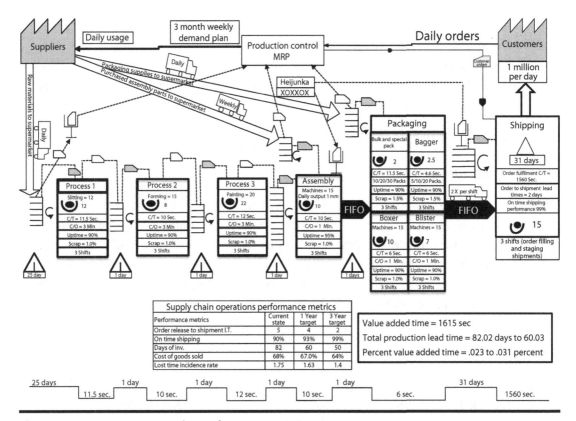

Figure 6.17 MMC Generation 1 future state VSM.

75,600 seconds/50,000 units = 1.512 seconds. A unit must be produced every 1.512 seconds to satisfy MMC customers. Takt time is used to synchronize all production operations to produce products at the daily customer average demand rate.

2. *Will a finished goods supermarket be set up or will orders be shipped directly to customers?* At MMC, there will be a finished goods supermarket as customer orders are filled from standard stock item inventory and shipped to customers.

3. *Where can you use continuous flow?* Because cycle times per unit are very similar between slitting, forming, painting, and assembly, it would be theoretically possible to consider continuous flow and have slitting be the pacemaker operation. But, since finished goods product is a combination of manufactured and purchased parts used in assembly to build the final product, it is preferable to establish assembly as the pacemaker and pull product from upstream processes. Continuous flow would be used through packaging operations, which has sufficient capacity and flexibility to deal with maximum assembly volume and product mix.

4. *What single point in the production process should be the pacemaker process?* Assembly is the logical pacemaker process choice as the final product is configured at this point from purchased and produced materials, therefore regulating their consumption.

5. *Where will MMC need to use supermarket replenishment systems?* MMC will utilize supermarket replenishment systems in seven locations starting with a

finished goods super market in shipping that will send production kanban cards to production control for scheduling assembly using a heijunka box to sequence products produced for packaging. Supermarkets will also be established between painting and assembly, forming and painting, and slitting and forming to support assembly production demand variability. In addition, three purchased material supermarkets will be established for supplier replenishment of packaging supplies, purchased assembly components, supplier replenishment of raw material, and metal coils for slitting.

6. *How should MMC level the pacemaker process mix?* To support volume mix requirements and facilitate efficient finished product handling, batch sizes are defined as ¼, ½, and full pallets of product. Finished product is packaged in 10 units per master carton and 100 master cartons per pallet with 20 master cartons per layer or 200 units per layer.

7. *What increment of work, called pitch, will be consistently released and taken away at the pacemaker process?* Pitch is the product quantity to be scheduled. Short scheduling cycles are important to greatly reduce lead time and enable level scheduling product mix to meet daily customer demand of 50,000 units or 5000 master cartons. Customer orders for 20 master cartons represent 60% of shipments. Schedules are released to assembly in increments of 100 master cartons, a full pallet of a single product or a full pallet equivalent made of more than one product in combinations of 40 master cartons, 2 pallet rows high and 60 master cartons, 3 pallet rows high, or 12.6 minutes of assembly time (100 cartons × 10 units/carton × 1.512 seconds/unit = 1512 seconds or 25.2 minutes).

8. *What process improvements will be necessary for value stream flow as defined by the future state?* Across the board, with the exception of bulk and bag packaging operations, changeover times, machine uptimes, and scrap need to be improved to initially achieve 90% uptime, 1% scrap, and changeover times of less than 3 minutes. These improvement opportunities will be implemented by projects defined in waves 3 and 4 through practices, such as 5S, process stabilization, work cell process redesign, single minute exchange of dies (SMED),[3] along with preventive and predictive maintenance, creating conditions for success in implementing system flow and pull in wave 5. In addition, the future state includes defining process target conditions, such as headcount, inventory levels, scrap percentages, machine uptime percentages, etc., the necessary inputs for estimating expected financial benefits of achieving 1-year targets illustrated in MMC's future state's supply chain operations performance metrics.

Step 3: Third, Future State, and Current State Maps Are Compared to Define Improvement Projects

Specific practices and tools are determined to support implementation projects to achieve the LEOMS's future state. For example, MMC's initial improvement

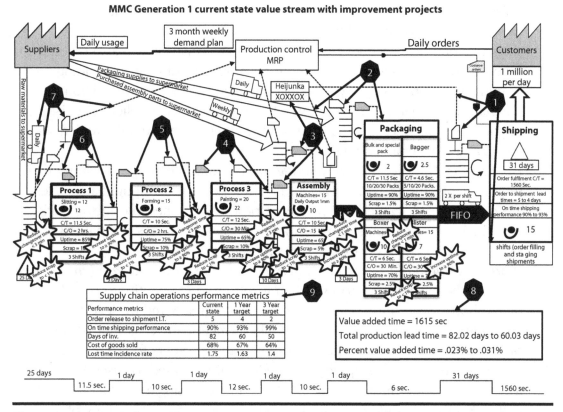

Figure 6.18   MMC Generation 1 current state value stream with improvement projects.

projects for each plant process area were identified and noted on the current state value stream map with black burst, creating Generation 1 current state value stream map with improvements (see Figure 6.18) for each process.

MMC Value Stream Improvement Projects

1. Shipping projects:
   a. Immediately increase shuttle frequency from packaging to shipping operations to twice per shifts, crew order filling and shipment staging three shifts, and schedule two outbound freight pickups per day, one at 7 a.m. and one at 7 p.m. These improvements combined with stabilizing, standardizing, and improving the finished goods order release through shipment process will smooth out flow and reduce process cycle time from 5 to 4 days, an improvement of 1 day (black hexagon 1, Figure 6.18).
   b. The largest contributor to preventing a 4-day order fulfillment cycle time is product availability failures. Current customer orders typically have 15 to 20 line items including some low volume products. Current scheduling and flow management processes result in stock outs, high levels of in-process inventories, long replenishment lead times, and constant "expediting" of replenishment orders. A finished goods supermarket will be established with replenishment kanban cards being sent to production

control who will schedule assembly using heijunka, creating a stable product flow and daily output to meet customer demand.

2. Optimize packaging operation (black hexagon 2 in Figure 6.18). Packaging operation cells will be organized into integrated cells focused on the four types of packaging machines: bagging, boxing, bulk packing, and blister packing, with operators moving from machine to machines to maximize machine throughput as assembled product flows into their cell. These changes will reduce operators on each shift providing the needed resources to staff material delivery routes and support continuous operations through lunch and breaks on all scheduled shifts.

   a. Elimination of lost production time because operators replenish their own materials, which can be accomplished by setting up a supermarket for all packaging area purchased packaging items and having team members from each shift reassigned to be material handlers to make scheduled deliveries and pickups multiple times during each shift.

   b. Applying FIFO flow into and out of packaging will replace packaging operation scheduling, as capacity is sufficient to process peak assembly volumes and mix. Since assembly is the pacemaker operation and sequenced based on highest priority needs, warehouse inventory levels are sustained to service customer orders on demand. FIFO replenishment of finished goods inventory from assembly ensures that scheduling sequence integrity established at assembly is maintained.

   c. Bulk and bagging operations have uptimes of 90% and scrap of 1.5%, which can be improved, but initially focus will be placed on boxing and blister packing that have 35% uptimes, 5% scrap, and changeover time of 30 minutes. These must be improved to 90% uptime, 1.5% or less scrap, and 1-minute changeover times.

   d. Boxing and blister pack machines have 35% uptimes and scrap rates between 1.5% and 2.5% from multiple causes that will be addressed during implementation of waves 3 and 4.

3. Assembly, the pacemaker operation, improvements (black hexagon 3 in Figure 6.18):

   a. In Lean pull systems, only one operation is scheduled by production control, called "pacemaker," as it sets the pace for other value stream processes. Finished assembled product will flow FIFO to packaging, as sufficient capacity is available to process peak volumes from assembly. The current 4 days of inventory between assembly and packaging currently will be reduced to 1 day. An important Lean solution for resolving this apparent dilemma is level scheduling; it involves three principle concepts: Takt time, small lot sizes, and a period of time, called pitch, to accumulate orders for leveling volume. MMC Takt Time = 1.512 seconds. Current daily mix and volume variability of order results are a significant challenge for scheduling to maintain level plant loading and meet customer demand as upstream processes are not reliable enough.

    b. Material will flow FIFO from assembly through packaging to shipping reducing inventory between assembly and shipping and enabling a finished goods inventory reduction in shipping.

    c. Supermarkets (planned inventory) will be added for purchased materials from suppliers. Supermarkets are replenished based on actual usage by consuming operations and displace use of MRP systems to drive purchase order issuing and rescheduling. MRP systems will continue to be used for forecasting material usage and capacity planning, but not for driving replenishment cycles.

4. Painting, forming, and slitting processes will be scheduled based on supermarket pull signals from the next downstream processes to replenish their supermarket (black hexagons 4, 5, and 6 in Figure 6.18).

    a. The PDCA improvement process initiated in wave 3 will be applied rigorously in wave 4 to identify root causes of quality defects and OEE including its components such as machine downtime and changeover time. SMED will be applied to optimize labor consumption, reduce safety risks, and increase uptime percentages.

    b. Waves 3 and 4 build a work cell operational foundation and process efficiency, and wave 5 connects the islands of value stream operations, creating end-to-end pull and flow to match customer demand rhythm. In MMC's case, replenishment supermarkets will be created between all operations except between assembly and packaging, where FIFO flow will be used as packaging has sufficient capacity to handle maximum volumes from assembly.

    c. Replenishment supermarkets (planned inventory) will be implemented in wave 5 to feed in-process operations except packaging. While material moves from packaging to finished goods using FIFO flow, a finished goods supermarket is established with replenishment kanban request sent to production control to schedule the pacemaker operation, assembly. Supermarkets are defined to have only the inventory need to cover supplying operations replenishment cycle time, and a level of safety stock. Supermarkets are used to support flow between two operations where factors such as lot size, cycle time, uptime, scrap, and changeover time are not suitable to support continuous flow. Supermarkets are not value-added but a necessary counter-measure until process stability, throughput, cycle time, uptime, and changeover times allow value streams to operate in true continuous flow.

    d. Material movements between production operations and purchased materials will be handled by designed material delivery routes to optimize material handling resources and provide consistent service to all work cells.

5. Production control will provide suppliers with daily raw material demand and a weekly demand forecast for the coming 3 months (black hexagon 7 in Figure 6.18). Inventory supermarkets will be established for raw materials supplied to process 1 and daily demand used to trigger delivery of more materials. Utilizing daily deliveries, inventory will be reduced to 25 days

from the current 30 days for raw materials, allowing the kanban to be created on the warehouse floor by establishing FIFO lanes for each of the 20 coiled materials products with cards from each day's usage given to production control to replenish inventory.

6. Expected improvement of process value-added time and lead times is shown at the bottom of Figure 6.18 with totals shown in the bottom right-hand corner (black hexagon 8).

7. Supply chain performance metrics improvement targets for year 1 and year 3 are shown in the MMC Generation 1 current state value stream improvement projects (black hexagon 9 in Figure 6.18). MMC could see that they have opportunities to meet and possibly exceed their year 1 and year 3 days of inventory targets as the total effect of identified improvement projects will result in reducing the current 82 days of inventory to at or below the 1- and 3-year targets of 60 and 50 days, respectively.

Initiate PFEP

PFEP, or plan for every part, is a data file related to each part from raw materials to finished goods that is vital to supermarket and kanban system implementation; they will be dealt with in waves 4 and 5. While some data should already exist on the item master database, there are other data specifically required to support implementation of waves 4 and 5, so it is advisable to start defining and documenting these data well before getting to the point where they are needed for implementation. The required data file is defined in *Making Materials Flow*,[4] p. 16. The IT department should also be engaged to modify item master database applications to allow PFEP unique information to be included as item master data along with an application allowing users easy access to PFEP data for items related to a supermarket they are creating. Early preparation of PFEP facilitates making choices such as container sizes and the number of containers required. Design of storage areas and storage systems to be used must be done well before needing them for implementation to prevent delays in implementation of supermarkets and pull systems in waves 4 or 5.

Define and Manage Projects Using A3 and Bowler

Projects should be divided into obvious "quick hits" and A3 projects. Quick hits are simple tasks that just need to be done as their solution is obvious and can be managed from a quick hit task list and done by existing work cell resources. Quick hits will get frontline operations team members and maintenance team members engaged, creating positive momentum by improving their work environment and daily results and energizing involved team members. A3s should be utilized for all significant value stream improvement projects identified (see Figure 6.19). The A3 process is an important tool for development of frontline team members as it is intended to be created as project documentation that is finalized through dialogue between mentors and mentees, typically a more

Wave 3 5S plant wide implementation	Dept/Business: MMC		1/4/16
	Team: All team members	Proj. Leader: Plant superintendent	
Title: 5S Plant wide implementation plan	**Recommendations:**		
Background: The foundation of LEOMS	Implement 5S supported by standwork instructions, PDCA, daily checklist for team members and layered audits administered by group leaders, team leaders, area managers, and plant operations superintendent.		
5S is the foundation step in Lean Operational System implementation. 5S creates an organized efficient workplace that improves safety, productivity, and team member ownership. It is the first step in establishing process stability, (the capability of a process to produce consistent product within a narrow range of variability around the specification target). This important step, improving capability through the consistent execution of processes, will be implemented throughout plant operations in every production, maintenance, quality, warehousing, and shipping area.			
	Plan:		
	All managers are responsible for implementation and sustaining 5S in their areas of responsibility. Their implementation team will include the area process engineer, maintenance technician, group leader(s), and team leaders. The project will include a 90 day initial implementation period follow by a 90 day period to establish sustainment daily routines.		
Current situation:			
Current work areas' organization and cleanliness vary greatly and little effort by the organization has been expended to support shop floor team member involvement in owning the organization and maintainance of their workplaces to optimize safety, quality, and efficiency.			
	Follow up:		
	Weekly the plant superintendent will review each process area's progress with the area manager and provide counsel and assistance as needed to keep the team on track. Progress charts will be maintained in each area showing completed implementation steps.		
Goal:			
Engage shop floor team members in continuous improvement of their assigned work stations to reduce waste, increase productivity, quality and work place safety.			
Analysis:			
Management has not established a system for organizing and sustaining safe, efficient workplaces throughout plant operations, but is now committed to establishing 5S as the plant process to be implemented and sustained by all plant operational and plant leadership team members.			

Figure 6.19 MMC improvement plan project A3.

experienced team leader, group leader, or superintendent and one of their team members. These dialogues are a valuable element of developing PDCA problem-solving competence as discussed in Mike Rother's book *Toyota Kata*[5] (page 220) and Art Smalley's book *Understanding A3 Thinking*.[6] A3 is used effectively for a number of purposes, including annual plan project deployment, project management, and problem solving. The content should

- Provide background, context, and importance of this problem
- Assess current performance and the gap vs. standard and perfection
- Determine specific goals and objectives of this project
- Analyze problems to identify their root causes
- Recommend countermeasures that will eliminate defined root causes
- Define action plan to implement recommended countermeasures
- Establish measurement, risk mitigation, a review process, and standardization plan
- Be reviewed and signed by the area manager

There are many benefits of organization-wide A3 use. Among them are the following:

- A3's standardized presentation format is understood organization-wide, so energy and focus are on solving problems.
- Increased speed of problem solving as repetitive use builds stronger problem-solving skills.
- Idea generation from the entire organization is encouraged as they are posted in areas accessible to everyone.
- Confirmation to frontline associates of their role as full team members with management.
- Joint A3 completion by frontline team members and leaders provides a great development experience and results in higher commit-ment to project objectives and time lines by those involved in its development.

Every process area department should have a glass wall to maintain wave implementation Bowlers, current performance metrics, audit results, audit trend charts, A3 implementation, and PDCA projects along with team member training certification status charts, training matrix board (to be covered later in this chapter), and open space that is sufficient to support team meetings, problem solving, and communications. Wave 2 completes direction setting by defining facility and value stream improvement opportunities and establishing value stream improvement plans to achieve established wave 1 targets. Waves 3 through 7 are a roadmap for change.

Wave 3

Building the Foundation: Wave 3

The purposes of this wave are to initiate establishing operational stability and to improve workplace flow, safety, and productivity. A basic premise of process improvement states that one cannot improve an unstable and disorganized process. Lean implementation's first step in achieving value stream stability is 5S implementation (see Figure 6.20). 5S is foundational as workplaces become more organized, safer, cleaner, and more efficient supported by standard work instructions developed for workstations; process steps, housekeeping tasks, quality checks, and connections to other stations are defined and documented. Often during this process, many ideas are generated to improve processes and workplace activities, and they should be documented on a list and kept for application during the appropriate implementation wave.

Frontline team members, process engineers, and team leaders work together to implement 5S (see Figure 6.21), a very important step as it initiates engagement of frontline team members in taking ownership for their workplace. The

Lean operational management system implementation dashboard		
For each target, enter your starting point and shade in the number of months to complete each wave. Monthly plot progress in achieving the milestones for each month. If progress has not achieved the milestones a corrective action plan is required to be implemented to get back on plan.		

Priority	Deliverables and measures	Starting point		1	2	3	4	5	6	7	8	9	10	11	12	13	14	15	16	17	18	19	20	21	22	23	24
1	Initiate culture change plan implmentation month 1		Plan																								
			Actual																								
2	Initiate wave 1 month 2		Plan																								
			Actual																								
3	Initiate wave 2 month 3		Plan																								
			Actual																								
4	Initiate wave 3 month 5		Plan																								
			Actual																								
5	Initiate wave 4 month 8		Plan																								
			Actual																								
6	Initiate wave 5 month 11		Plan																								
			Actual																								
7	Initiate wave 6 month 12		Plan																								
			Actual																								
8	Initiate wave 7 for annual strategic planning process		Plan																								
			Actual																								
9	Create sufficient improvement cost benefits to fund year 1 startup cost		Plan																								
			Actual																								

Figure 6.20   LEOMS Bowler wave 3.

Create foundation wave deliverables:
- Key personnel trained on 5S
- 5S implementation plan
- Clean and organized workplaces
- 5S standard work procedures
- Introduce Kaizen PDCA
- 5S checklist for all operations
- Management checklist to support improvement and sustainability
- Initiate suggestion system
- Quick hit facility and equipment improvement projects
- 5S facility scorecard and measurement process
- Three year facility flow plan
- Reduced wasted effort and safety hazards
- Glass walls established for the plant and each processing area/department

Information required
- VSM current and future state with current and 3 year volumes (product defined at the level (stock number, product sub-family, product family) necessary to identify what will be required from the manufacturing process)
- Desired performance of the key business and operational metrics
- Product volume forecast – finished goods-in process-purchased materials

Figure 6.21   LEOMS wave 3 plant implementation.

previously discussed training method (refer to Figure 6.9) should be applied to train frontline team members and initiate transferring ownership responsibility to them. Seeds of ownership need to be firmly implanted and nurtured at every step of this journey during implementation, as this is a vital activity in establishing an effective LEOMS culture. 5S implementation is managed using Bowler to keep all departments moving forward in synchrony. 5S implementation dashboard (see Figure 6.22) is a model that should be used to track and manage implementation of all seven waves and posted in each area and on the

Lean Operational Management System 5S Implementation Dashboard																		
For each target, enter your starting point and shade in the number of months to complete each wave. Monthly plot progress in achieving the milestones for each month. If progress has not achieved the milestones a corrective action plan is required to be implemented to get back on plan.																		
Project Status	Last Status Review Date:			Green	On-time	Yellow	At risk	Red	Off-plan		Initiate		Continue					
Priority	Deliverables and measures	Starting point		Month	1	2	3	4	5	6	7	8	9	10	11	12		
1	Shipping			Plan														
				Actual					On-time									
2	Packaging			Plan														
				Actual					On-time									
3	Assembly			Plan														
				Actual					On-time									
4	Painting			Plan														
				Actual					On-time									
5	Forming			Plan														
				Actual					On-time									
6	Coil slitting			Plan														
				Actual					On-time									
7	Receiving			Plan														
				Actual					On-time									
8	Maintenance shop			Plan														
				Actual					On-time									
9	QA			Plan														
				Actual					On-time									

Figure 6.22   LEOMS department 5S implementation.

plant glass wall. Implementation of 5S (sort, set in order, shine, standardize, sustain) across every operation creates the first step in establishing real process stability. The purchase of *5S for Operators*[7] from Productivity Press is highly recommended as an excellent training booklet that includes teaching fundamental motion economy and safety principles. 5S is often thought of as housekeeping, and while it does involve housekeeping, its purpose is creation of stability and team member ownership. Stability, in this context, means quality product produced defect free, consistent process variability over time, and within a consistent cycle time (see Figure 6.23).

Continuous Improvement through Problem Solving

Creating an army of problem-solving scientists is achieved in the frontline cauldron where defects and variation are continually bubbling up and always will be for two reasons: First are the continuous consequences of the second law of thermodynamics, "left alone all processes will deteriorate"; and second, perfection will never be achieved as "perfection conditions" will continually be raised. But the good news is that this creates a great place to develop a highly effective army of problem solvers as they hone their PDCA process skills. See Figure 6.24 to understand the correct steps and their application sequence. Learning to apply PDCA can be done by following the five elements as shown in Figure 6.25. The first steps of the process are "how" questions, which lead to development of potential cause theories, followed by five-why cause investigations leading to identifying the real root cause (see Figure 6.26). This is often a process that is easy to understand but takes rigorous practice to become proficient. PDCA is

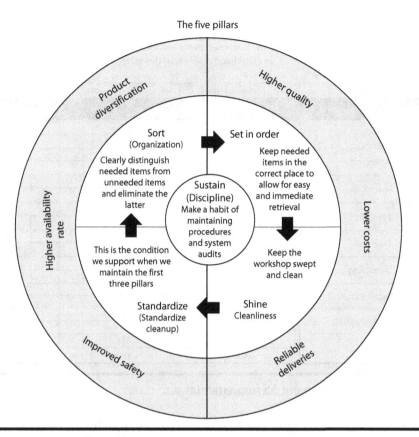

Figure 6.23   LEOMS 5S five pillars.

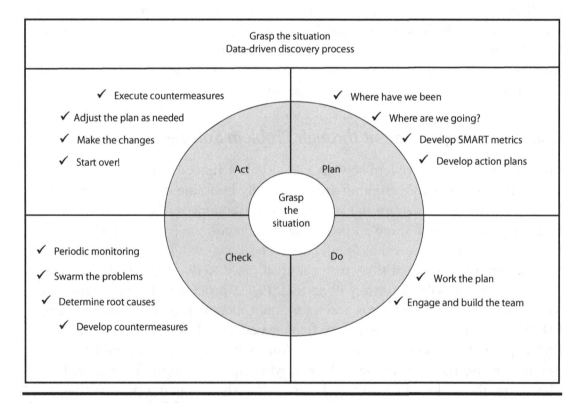

Figure 6.24   PDCA model.

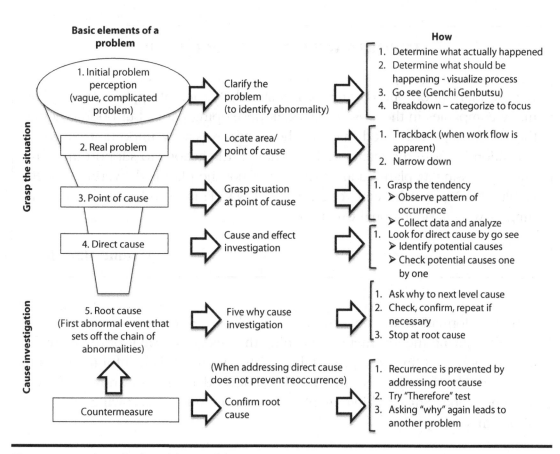

Figure 6.25 Practical problem solving.

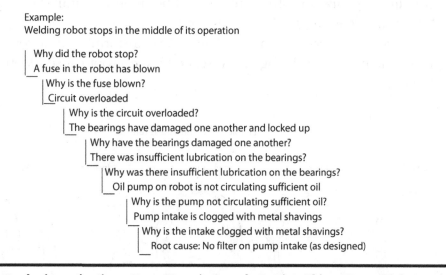

Figure 6.26 5 why investigation. (From Dennis Pascal, Getting Things Done Right.)

something that will only be learned by students when coaches by making the student "suffer" to find the answers.

Another valuable approach in problem solving is learning from data by understanding how data can be transformed into knowledge, and ultimately predictability (see Figure 6.27) as raw data moves through five steps starting with data

---

IMPORTANCE OF THE CHECK STEP

I can recall during my time at NUMMI where I embraced PDCA, while walking the plant floor, one day my sensei stated, "Jerome it appears to me that many companies in the west are good at developing plans, but poor at doing the check step." He stated that the check step was very important for good execution because the team needs to know if the actions to support the plan are progressing as planned and whether or not the plan is delivering desired results. PDCA must be embedded in our DNA. This includes business planning as well as operational execution.

Jerome Hamilton

---

leading to information, the understanding of relevance or context leading to knowledge, giving the information meaning that leads to insight or the nature of information, and finally to predictability. This approach is helpful in efforts to understand why something works, which is a level deeper than knowing how something works. Shigeo Shingo explains the importance of knowing why.

Problem solving tips:

1. When visible abnormalities appear, "swarm the problem"; this infers that frontline organization structures should be designed similar to Toyota's, which allows group leaders to respond when needed.
2. Favor actual observation of problems vs. brainstorming in a conference room; nothing beats being at the scene to keep root-cause identification focused on reality.

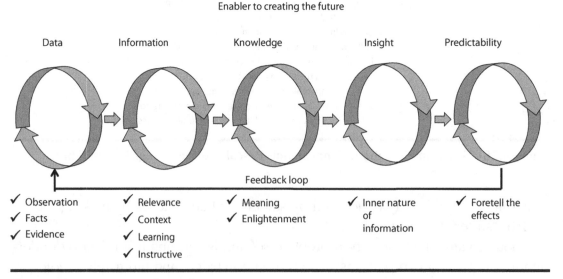

Figure 6.27 Data transformation.

"With know-how, you can operate the system, but you won't know what to do should you encounter problems under changed conditions. With know-why, you can understand why you have to do what you are doing and hence will be able to cope with changing situations."

Shigeo Shingo

3. Perform rapid experimentation to implement temporary countermeasures until permanent countermeasures/solutions are found and implemented.
4. Stop/fix problems within Takt time to not cause lost time, missed shipments, and rescheduling. Violating Takt time has a major cost penalty as all plant cost will not be "absorbed" by a unit of product during lost time, resulting in higher fixed and variable unit cost.
5. Take determined action to reach toward achieving target state conditions, as it is the only way to continuously improve quality and reduce defects/cost.
6. Practice knowledge sharing—yokoten:
   a. Horizontal, peer-to-peer sharing of what, how, and replication of improvements
   b. Systematic dissemination
   c. Knowing "why" is as important as "how"
   d. Sharing methods: 5 minute meetings, team/group meetings, storyboards, and/or learning labs

The enormous challenge of building an army of scientists solving problems everyday cannot be overstated; it is key to building a genuine continuous improvement culture. Problem-solving scientists are built with daily problem-solving routines over time. Kaizen workshops have their place, especially in establishing the LEOMS and making system level improvements, but they are not the principle engine to building a strong foundation of continuous improvement culture.

Consistent coaching and repetition is the only known path to building an army of problem-solving scientists who will master "imbedded routines" and over time develop into highly skilled problem solvers, who like a skilled craftsman get better and faster with practice.

Practical Benefits of 5S

- Workplaces are organized with a place for everything and everything in its place making the workstation more productive.
- Improved safety with nothing to trip on or lift in an unsafe body position.
- Fewer workplace injuries as principles of motion economy are applied resulting in ergonomically safe workstations requiring less stress, strain, and effort.

KAIZEN WORKSHOPS

Improvement workshops are special improvement efforts that temporarily bring together a team of people to focus on a particular process. The duration of a workshop is typically one to five days. Workshops are used extensively and do have their place. Toyota utilizes workshops too, for example, but not as its primary means of continuous improvement.

Mike Rother
Workshops vs. Kaizen, Toyota Kata, p. 25

CONTINUOUS IMPROVEMENT "IMBEDDED ROUTINES"

Toyota's superior results spring more from routines of continuous improvement via experimentation than from the tools and practices that benchmarkers have seen. Many of the tools and practices are, in fact, countermeasures developed out of Toyota's continuous improvement routines. We won't be successful in the Toyota style until we adopt a do-it-yourself problem solving mode.

Mike Rother
Toyota Kata, p. 7

- No loose dirt or debris to contaminate product or damage equipment.
- Improved equipment life as oil leaks, rust, or filings are quickly identified and the source of the problem is fixed.
- 5S also introduces PDCA and begins building the army of problem solvers.

5S work instructions are documented in standard work documents and sustained by daily and weekly checklist and audit systems. Responsibilities for maintaining, validating, and verifying daily team member compliance documentation are assigned to appropriate leadership levels from frontline team members, group leaders, area group leaders, team leaders, to department superintendents. These processes are established to build sustainability and continuous improvement into 5S, ensuring stability through effective maintenance and continuous workplace improvement. 5S is the first step in gaining value stream stability and needs to be thought of like the foundation of a house: it must be strong and reliable.

Engagement Starts Now—Start Creating the Ninjutsu

5S is the initial engagement of frontline associates in taking ownership for their workplace, so previously discussed training methods should be applied to train frontline team members including transferring 5S process ownership responsibility to frontline team member trainees. The seed of ownership needs to be firmly implanted and nurtured along the way during LEOMS implementation. From day 1 of 5S implementation, leaders must demonstrate their people management coaching skills as the MMC way leaders interact with first line team members. It is a great time to implement a facility suggestion system, encouraging frontline team member involvement. No one knows day-to-day operations better than team members doing the job every day. Building this "we" team paradigm with frontline team members as co-owners of their work area will contribute to increased job security and satisfaction of team members as they participate as co-decision makers. Most frontline team members have wanted to have opportunities to see their ideas implemented, so it is a great opportunity for leaders to start work cell redesign. Leaders should also take this opportunity to practice their responsibilities as teachers and coaches (see Figure 6.28). Creating management magic through building a continuous improvement culture is leadership's most difficult and most important contribution to their company becoming a Lean enterprise. 5S training and implementation should focus on transferring principles and practices, charging frontline team members with responsibility for implementation in their areas. Team members should be expected to absorb and follow the 5S training in transforming their workplace as they try out their ideas, brainstorm with their colleagues, and continuously look for improvement opportunities. This approach will require iterations of assessment and coaching, but it is the starting point of creating the right genes, enabling organizational development of a strong continuous improvement culture. This will take effort from engineers and supervisors to facilitate and train team members to fully participate in work area and station redesign while implementing 5S. This is only the first step to achieving long-term frontline team member ownership and development as problem solvers and owners of their assigned process. They need to become "industrial engineers and safety engineers" by learning basic industrial engineering and safety engineering workplace design principles to enable them to take ownership for redesigning and sustaining their workplace. Frontline

The LEOMS' leader people leadership expectations:

1. Engages people at all organizational levels

2. Teaches team members to focus on the work, material flow, and value stream to see the waste

3. Gives team members deep technical and process knowledge

4. Pushes responsibility for value stream management and improvement to the lowest possible line management level

5. Introduces metrics to encourage horizontal thinking

6. Creates frequent problem solving loops between themselves and their superior and themselves and their subordinates

7. Accomplishes these six leadership responsibilities through application of policy deployment, A3 analysis, standardized work with standardized management, and kaizen

Figure 6.28 Lean leader's people leadership.

team members will only begin to see themselves as part of management when they start to make final decisions and own the resulting performance. Frontline team members must see themselves as having meaningful involvement in decisions related to continuous improvement in their work area. 5S is a first small but important step in development and application of management magic or Ninjutsu (see Figure 6.29).

Waves 4, 5, and 6 complete the content to be mastered until, like a great magician, suddenly visible manifestations from team members on the frontline appear simple, efficient, purposeful, effective, and consistent. Achieving this state requires persistent dedication to practicing the "art" as a magician must do, until the magic is observable and consistent. Lean implementation uses both top-down and bottom-up approaches to deployment. Planning and alignment happen from top down to assure that achievement of operational performance aligns with market and financial requirements and goals, ensuring improvement of the businesses' value proposition and strategy. Implementation goes from bottom up, and analogous to building a house, a strong foundation is required. We may have a great architectural plan that includes all our desires and dreams, but when it is time to build it, we start with the foundation. It is

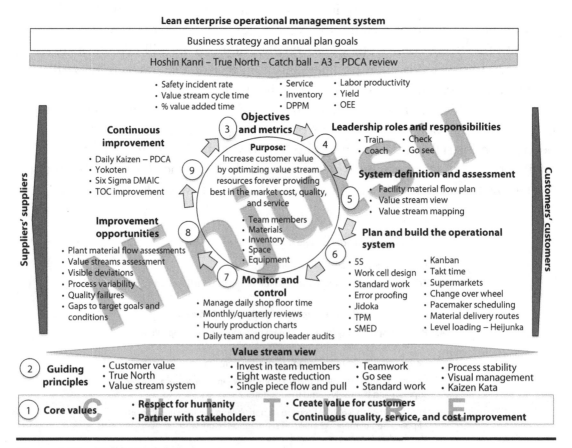

Figure 6.29 LEOMS Ninjutsu.

advisable to maintain close observation and verification of 5S activities until the organization adopts the process and it becomes a daily routine that will be sustained forever as shown in the 5S continuous circle of improving workplace stability (Figure 6.23).

5S Scorecard

A visible 5S scorecard should be established and maintained on glass walls related to audit performance along with active A3 projects and "just do it" documents. A history binder should be kept of all the completed projects, as an occasional review by operators reinforces where they have come from in improving their work area and occasionally to re-implement something that has slipped through the cracks.

Set Up Glass Wall Areas

Glass wall communication areas should be set up for the overall plant and each processing area or department. Glass wall areas maintain information related to performance and progress of LEOMS implementation along with in-process A3 projects and relevant communication information. Glass wall areas should be designed to support 5-minute start of shift meetings, group problem solving, and plant tours. Having the 5S established, it is time to move on to wave 4, work cell redesign (see Figure 6.30), the next step in constructing the Lean enterprise operational system.

Lean operational management system implementation dashboard																											
For each target, enter your starting point and shade in the number of months to complete each wave. Monthly plot progress in achieving the milestones for each month. If progress has not achieved the milestones a corrective action plan is required to be implemented to get back on plan.																											
Priority	Deliverables and measures	Starting point		1	2	3	4	5	6	7	8	9	10	11	12	13	14	15	16	17	19	20	21	22	23	24
1	Initiate culture change plan implementation month 1		Plan																								
			Actual																								
2	Initiate wave 1 month 2		Plan																								
			Actual																								
3	Initiate wave 2 month 3		Plan																								
			Actual																								
4	Initiate wave 3 month 5		Plan																								
			Actual																								
5	Initiate wave 4 month 8		Plan																								
			Actual																								
6	Initiate wave 5 month 11		Plan																								
			Actual																								
7	Initiate wave 6 month 12		Plan																								
			Actual																								
8	Initiate wave 7 for annual strategic planning process		Plan																								
			Actual																								
9	Create sufficient improvement cost benefits to fund year 1 startup cost		Plan																								
			Actual																								

Figure 6.30   LEOMS Bowler wave 4.

Wave 4

Redesigning Work Cells: Wave 4

Work cells are areas of activities in production, receiving, shipping, warehousing, material replenishment, order filling, maintenance, production control, quality control, and other plant functions. The deliverables of wave 4 are shown in Figure 6.31 and will result in improvements described below:

1. Reduced non-value-added work caused by current work cell designs and standard work instructions.
2. Redesigned work cells to create flow and synchronization of work among work cell team members eliminating non-value-added work, minimizing necessary non-value-added work, and optimizing value-added tasks.
3. Preventive and predictive maintenance methods applied to all applicable situations to minimize the disruptions of breakdown maintenance. Maintenance standard work is created, and the maximum routine maintenance activities are transferred to frontline associates to allow maintenance team members to focus their time on prevention of failures.
4. Redesigned purchased material flow to optimize inventory and improve availability through use of kanban and supermarkets, material replenishment delivery, and pickup route cycles, along with workstation part presentation to improve operator efficiency and application of ergonomics principles to improve safety.

Work cell redesign wave deliverables:
- Optimized work cell flow design
- Standard work instructions for all key tasks including changeovers
- Optimized and balanced work among cell operators
- Optimized processes with established standards for process set points, ranges, and corrective action for out of control conditions
- Quality at the source
- Total productive maintenance program
- Inventory supermarkets and kanban system replenishment for purchased materials
- Error proofing analysis and implementation of identified solutions
- Scheduled cell material delivery and pickup
- Day by hour chart implemented
- Operators and group leaders and supervisors apply PDCA

Information required
- VSM current and future state with current and 3 year volumes (product defined at the level necessary to identify what will be required from the process)
- Takt time
- PFEP plan for every part
- OEE
- 1 year and 3 year target performance of the key strategic and operational metrics
- Three year facility flow plan

Figure 6.31 LEOMS wave 4.

5. Reduced quality and productivity loss by applying error proofing and visible test to work cell tasks.
6. Redesigned work procedures documented as standard work instructions.
7. Operational day by hour charts to measure process output performance and Takt time compliance along with andon signal systems to call for assistance when needed to resolve issues interfering with Takt time compliance.
8. Develop a Pareto chart to explain 80% of the gap between current OEE and 100%, and prioritize improvement projects to identify and resolve the causes.
9. Introduce and begin building cell team member competencies to apply PDCA, continuously improving safety, labor productivity, work cell throughput, and quality.

Work Cell Redesign

A good training source for optimizing work cell operator value-added time is *Creating Continuous Flow* by Mike Rother and Rick Harris[8]; this common-sense easy-to-follow workbook takes readers through a process to optimize operator value-added time in work cells. In process operations, operators are frequently monitoring machines, a completely non-value-added activity. This can be a result of insufficient process control or a machine centric process design without much thought to operator activities and the resulting productivity loss and waste of their talent and knowledge. Solving process instability takes significant effort and skilled process engineers who have statistical process optimization study experience to better define and control critical process variables. Covering this topic is beyond the scope of this book; if you are interested in process optimization, become a member of the Society of Manufacturing Engineers (SME) or Learn how to apply the Six Sigma methodology. SME has programs, training, and reference materials for their members.

The work cell redesign is accomplished through analysis and redesign. Analysis activities include the following:

1. Apply value stream Takt time to establish work cell production rates required to satisfy customer demand rate.
2. Complete work sampling studies to quantify value-added time percentages, and identify occurrences and total time of non-value-added and non-value-added but necessary tasks being done by operators. This is a great activity to involve frontline team members to develop their "eyes" to see waste and take ownership for their process and work cell design.
3. Map the work cell and operator movement and apply the spaghetti diagram process to a work cell sketch; be sure to mark the frequency of each path traveled by operators.
4. Document all operator tasks, including walking, to create current state using standard worksheets or combination worksheets if machines are operated as a part of the cycle time.

5. Redesign work task and cell arrangement to eliminate waste in waiting, walking, and handling and reduce ergonomic risks.

6. Apply SMED to reduce setup and changeover times.

7. Utilize an operator balance chart as required when multiple operators are involved in a cell to perform analysis and redesign and distribute operator tasks to balance work among cell workstations.

8. Define materials and supplies required to complete work cell operations along with storage and parts presentation to minimize wasted operator movement.

9. Define material quantities to be stored at workstations based on replenishment cycle times of the material handlers to cover material delivery and pickup time plus safety stock to reflect the average wait time until they can respond to the replenishment signal. Building in-process inventory supermarkets, kanban systems, and fully integrated material delivery routes is part of wave 5, as it is best done when all the work cells have been redesigned and the system is prepared for pacemaker scheduling.

10. Revise current maintenance practices including preventive and predictive maintenance practices, ensuring that they are sufficient and effective along with maximizing the transfer of routine tasks to frontline team members.

11. Write/update standard work instructions for all operations, maintenance, and materials replenishment tasks.

12. Establish day by hour charts to measure process output performance and Takt time compliance.

13. Continue to reinforce team member PDCA problem solving so they utilize continuous improvement daily in their work cell.

Achieving success in applying the LEOMS requires that organizations possess and apply four essential capabilities as defined by Bluegrass Automotive Manufacturers Associates in collaboration with TEMA and Dr. Steven Spear in his book *The High Velocity Edge.*[9]

1. Process design
2. Problem solving
3. Sharing knowledge
4. Leaders as teacher

Effective Process Design

Effective process design requires four elements:

1. System output requirements (see Figure 6.32)
   1a. Processes must have defined prespecifications, the target level of performance, for example; output in parts per available hour equal to Takt time. (Remember, targets are set to drive performance toward perfection.)

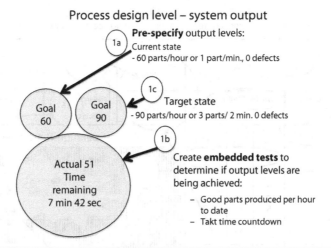

Figure 6.32   Process design, output.

1b. Processes must have embedded tests, for example; good parts per hour this shift and day or Takt time countdown. Failure detection based on current performance measurement needs to be "real-time" detection followed by PDCA problem solving and temporary and/or permanent countermeasures.

1c. Processes must have a defined ideal state (True North target conditions), and the currently attainable state on the path toward "True North." It could be thought of as an intermediate state analogous to intermediate future states in an iterative process toward the idealized end state.

2. Activities are the methods used to complete a required set of tasks. Standard work combination sheets are used to see value-added work and waste in the current process and layout design. This should be completed with involvement of frontline team members. Processes with both automated or semi-automated machines involved should utilize standard work combination sheets as shown in the example (Figure 6.33).

1a. Date document was created.

1b. Department and work cell.

1c. Work element.

1d. Element times: manual work time, machine time, and walk/wait time.

1e. Current standard quota per shift and Takt time.

1f. Target quota per shift and target Takt time

2a. Layered audits conducted by assigned levels of management verifying the conformance with the standard work.

2b. Team leader standard work audits.

3. Process pathway. Pathway defines the order of each activity and who is responsible; Figure 6.34 shows the nine process steps divided between three operators. Using a completed standard work combination sheet, process work elements are divided among operators by balancing work, walk, and wait time as evenly as possible among three operators in order

Process design level – work elements or activities
Pre-specify the work content, sequence, timing, location, and output.

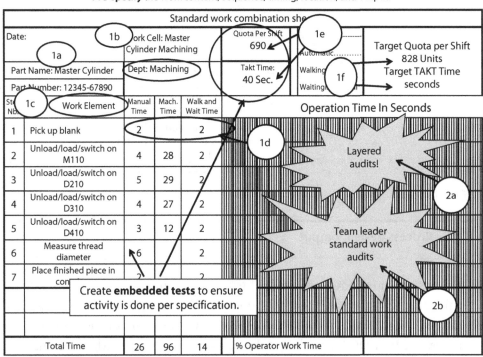

Figure 6.33 Process design, work elements, or activities.

1. Defines the order of process steps and the operator responsible for each step
2. Team leader observes who is performing activities per specification

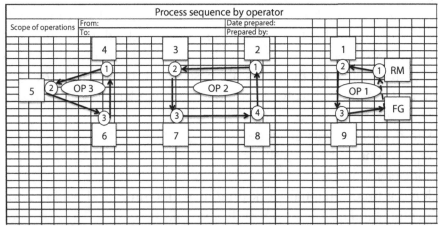

Figure 6.34   Process design pathway.

to minimize "built in waste" that occurs when their assigned times of operations are not equal. The second pathway tool is the training matrix board; Figure 6.35 shows which operators are trained on which processes, allowing team leaders to verify that process activities are being performed by a trained operator. The details of these documents are a critical part of

Embedded tests:
Determine ways to ensure activities are being performed by the trained person

Task Nbr.	Jackson	Jones	Baker	Henley	Smith
1					
2					
3					
4					
5					
6					
7					
8					
9					

Figure 6.35 Training matrix board.

maintaining standard work resulting in efficient operations, sustaining gains that have been made, making it easy to train new frontline team members and cross-train current team members on various work cells and processes. Because these documents are standardized across all operations in a plant, everyday occurrences such as covering for vacation, sick days, and employee turnover are handled with a minimum amount of effort and an assurance that productivity and quality will not suffer because a key person is absent.

4. Connection (handoff). The handoff makes connections between processes and their need for services or materials. As shown in Figure 6.36, process design connections are achieved through various communication request methods and when appropriate, used to escalate the urgency for providing materials or services. Typically, we think of andon cords as a tool to call for assistance in resolving problems and, after a specified time, to signal line

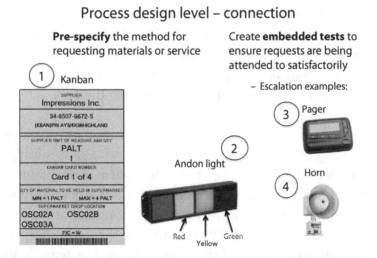

Figure 6.36 Process design connection.

stoppage. Hand-off signals also includes kanban cards, which are typically used to request production materials and supply items. Other common signal methods are:

a. Andon lights can simply indicate that a service is needed, or red, yellow, and green light showing their status: green means everything is running well, yellow indicates the service is needed, and red means the service is needed soon or the line will go down.

b. Pagers for verbal communications.

c. Horns and speakers that often play a specific song that indicates a specific problem or service is required.

In most non-Lean plants, management would be unhappy if their andon cords were being pulled regularly as they would see it as interrupting production and driving up cost. At Toyota, they see andon signals differently; they are opportunities to "invest time in building talent" in making processes more reliable and receiving a return as cost will be lower and quality will be improved. At one Toyota plant, team members typically pull the andon cord 12 times per shift on average, and in final assembly, it is pulled on average once every 1.25 minutes.[10] The great majority of these occurrences do not result in a line stoppage as all available assembly area team and group leaders "swarm" problems and assist team members in resolving problems quickly. The "swarm teams" are so experienced that rarely is Takt time violated.

Key points to remember in designing processes:

1. Outcomes must align with "True North."
2. Design work, equipment, and facilities to see abnormalities, find their root cause and take corrective action.
3. Three common process design failures:
   a. Lack of specifications
   b. Lack of embedded tests or countermeasures
   c. Lack of specifications of the ideal state
4. Define and execute validation that team members are performing their activities per specification; this is frequently done by maintaining a training matrix board or chart defining who has been trained for each task and who is currently assigned responsibility for executing the task (see Figures 6.34 and 6.35).
5. Process design connections are involved between steps in a process and at points of integration with support processes such as supplying materials, maintenance support, or assistance to swarm a problem. Requesting material with kanban cards is an example of a connection with another process. Other methods could involve using a pager, andon light, horn, or music from a loudspeaker. They can serve the same purpose, signaling needs for material, group leader support, and/or team leader to assist

the front line team member in maintaining product flow to achieve daily output targets. A signal may also be used to indicate that a need is urgent, allowing sources to prioritize requested responses to not interrupt line throughput.

The Natural Law of Waste

The natural law of waste is a very logical principle (see Figure 6.37); if a product that passes through sequential steps has a defect, due to excessive variation in the initial process step, it creates processing variation and potentially defects in downstream processes. As products pass through each downstream process, the cost of the initial process defect continually increases as more labor and machine time accumulates. When solving a defect at a downstream operation, finding its root cause can be very complicated if, in fact, it results from an operation three or four processes upstream. The point is made by an old adage, "What starts well, ends well." This also suggests that a good strategy for sustainably improving cost and quality of a product is to start by optimizing initial process steps and work downstream through the remaining processes. This will avoid spending hours improving a downstream process only to find out that the true root cause "seed" is upstream a few process steps.

Creating a "Run to Standard and Target" Culture

Discipline in Process

Run to Standard—Maintain the Gains

Run to standard is about holding gains and controlling process averages and variability to achieve consistent quality and productivity. It is achieved by rigorously following standard work by being vigilant in detecting defects and responding

Figure 6.37 Natural law of waste.

Management by process requires ongoing inspection and enforcement to ensure
"Standardized work" is being followed.

```
                    Standards

  Enforcement  ◄────────────  Inspection
```

Processes do not improve automatically in fact they tend to follow the second law of
thermodynamics – All systems degenerate when left to themselves

```
                    Standards

  Problem solve    Improve      Stabilize

              Identify waste
```

Figure 6.38 Discipline in process.

with countermeasures following PDCA (see Figure 6.38). The point is that processes will naturally deteriorate, meaning over time they will generate waste, from both familiar and new causes. Countering these effects must be managed through consistent discipline in operating and maintaining processes to specific standards so waste can be quickly identified and resolved.

Run to Target

Using the term "target" in this case does not only mean what we normally think of as a target; in Lean, it also means target condition(s), a description of how processes should operate to achieve desired outcomes (the target). Target conditions are a set of conditions or a pattern of activities that will generate desired outcomes, target performance. There is logic to this approach; for example, if I have an inventory reduction target and ask five colleagues to reduce inventory, it is likely that they will have different ideas about what actions to take to achieve it and may end up in conflict as they defend their ideas about what should be done. On the other hand, if there is a predefined pattern of activities or parameters creating conditions that have proven to reduce inventories, it is more productive to have teams focus on executing and improving this pattern of activities and parameters establishing conditions resulting in reduced inventory. A simple example is inventory levels; it is controlled by variables lead time, lead-time variability, demand, and demand variability. Lead time can be normally broken down into several segments, purchase order preparation, vendor, shipping, production, etc. Each lead-time segment has a historical average time and variability that can be broken down further to understand current activities/conditions that are "controlling" lead times. By focusing on changing these key direct variables/conditions, total lead times will change. We have all watched as finance issued an inventory reduction edict resulting in negative service and cost outcomes. Instead, by having team members involved in the workplace apply the LEOMS's very logical "run to target" approach, they stay focused on maintaining performance by following current standard conditions and making improvements to reach target conditions resulting in higher performance and new standard conditions. Once we grasp this idea, it becomes "obvious" to us because it is logical that by identifying and improving key direct variables, outcomes will improve.

Metrics   Lean uses metrics to measure results and process effectiveness. Hoshin Kanri planning ensures selection of the most important metrics in alignment with key customer and business priorities. Frontline metrics are linked to top-level annual metric targets to improve as the catch ball process and tiered Bowler processes are applied all the way to the frontline, thus creating ownership at each organizational level.

Monitoring and Control   The LEOMS is a people- and culture-based system, so monitoring and control rely on managements' frequent frontline presence. Visual management is the LEOMS's sensing mechanism, providing transparency of operational reality and data related to improvement. It also provides clarity of deviations against detailed standards of performance, work procedures, schedules, inventory levels, and scrap percentages, among many others. For example, inventory in a supermarket should never exceed defined maximum levels. Storage locations in supermarkets are designed to hold the maximum planned inventory. If inventory exceeds the maximum, it must be stored outside designated locations creating a visible deviation that will be easily observed. These easily observed deviations demand immediate follow-up to understand root causes of system failures. Another example is daily review of "day by hour" production charts showing output achieved each hour vs. planned rate. These charts include identification of deviations, both positive and negative, definition of their root causes, and an explanation of actions taken to permanently resolve negative problems. Positive deviations should also be examined for their root causes as this might lead to understanding how to permanently improve process performance. All shop floor functions are accountable for responding quickly to frontline issues so they are resolved with minimum flow disruption. The second key element of Lean monitoring and control are tiered audits of standard work. Daily, each organizational level audits a specific area of standard work. During a plant manager's daily frontline walks, audits of standard work are performed on standard work previously audited by area managers. This completes a cycle of audits, which started with frontline team members followed by team leaders, group leaders, area superintendents, and finally the plant manager.

1. The first result of these audits is coaching area managers, supervisors, lead people, and frontline associates about the LEOMS through keeping data on noncompliance. The LEOMS's audits are retained and summarized in a

matrix that includes all items audited in one manager's operation on one axis and auditors' names on the other axis. Ratings are discussed with the entire audit team to maintain calibration between raters and gather audit system improvement ideas.

2. A second summary is made by audit topics across all facility operations. This allows management teams to see common weaknesses. Value stream stability depends on standardized operational discipline; audits are control mechanisms for maintaining system discipline. Lean audits are serious business and are included in annual reviews of every frontline team member, team leader, group leader, and area superintendent.

Plant manager daily walks build a healthy tension of joint accountability between frontline team members and managers. If problems occur repeatedly at a given frontline operation, team members will look to managers to engage and assist them by coaching them through the process of resolving these more systemic problems. This creates a healthy tension and reinforces a positive work environment as managers convey support and respect for frontline team members. As an operation's Lean system matures, a more standardized problem–response process follows a well defined escalation path, ending with engagement of managers when problems are not resolved within a predetermined time. The plant manager's daily walks include auditing operations to keep frontline team members, their group and area leaders accountable for executing their roles and coaching them.

Improvement Opportunity Identification and Problem Solving   In summary, the LEOMS's tools for identifying improvement opportunities are VSM, audits, and visual management. VSM is the system-level process tool, where management is responsible for defining and implementing improvements. Frontline visual management drives long-term continuous improvement. Daily deviations are exposed through visual management and audits managed by frontline teams as they are responsible for identifying root causes of problems and implementing permanent solution through application of the five why's. First, ask why a problem is occurring. The response is then met with a second "why is this occurring?" This process continues with each successive response met with another "why" until a root cause is determined. Through experience, practitioners have learned that it normally takes up to five cycles to locate the real root cause. After root cause identification, "plan do check act" (PDCA), a four-step improvement process, is applied. This simple method is based on the scientific method taught by Edward Deming.

These simple standardized tools allow frontline team members to learn and apply them. Improvement methodologies, such as Six Sigma DMAIC and TOC, are not well suited for everyday use by frontline team members as they are more complicated and require significant expert knowledge.

THE FIVE WHYS

The Toyota Production System has been built on the practice and evolution of this scientific approach. By asking why five times and answering it each time, we can get to the real cause of he problem, which is often hidden behind more obvious symptoms. In a production operation data is highly regarded, but I consider facts to be even more important

Taiichi Ohno
Toyota Production System, Productivity Press, 1988

The Lean Practices   During its evolution, the Toyota Production System developed myriad practices to expose and solve problems resulting in elimination of waste and flow improvement. In addition to continually improving value stream designs and elimination of deviations, Lean also employs Hoshin Kanri, an operational planning approach ensuring deployment of only those resources required to support the rate of leveled customer demand. The LEOMS also establishes customer demand rate using a method called Takt time, discussed earlier in this chapter. These practices are pragmatic methods of ensuring that resources are planned and consumed effectively and efficiently and sustained through standardized work practices enabling control of consistency, productivity, and rapid defect resolution.

   This array of practices is applied to value streams as appropriate through each continuous improvement cycle. Lean tools are a means to building, sustaining, and improving a company's LEOMS.

Roles and Processes   Lean standardized work defines the LEOMS's roles for every person in an organization, including management. Implementation normally starts on the frontline. Frontline job activities are classified in three categories: value-added, non-value-added (waste), and non-value-added but required. For example, if product assembly process is analyzed, handling of parts is waste,

STANDARDIZED WORK AT ALL LEVELS

The Introduction and early deployment of LEOMS within 3M included training 3M manufacturing leadership as Rick Harris conducted operational reviews with all 60 US facilities in North America with 6 to 10 facility managers at the sessions held every four to six months. During one of these sessions he trained the participants on Standardized Work for Managers.

actual assembly is value-added, and quality checks are required non-value-added work. Work is organized in cells with task balanced among cell operators to ensure high utilization of each operator. Each job is analyzed to eliminate non-value-added work, minimize non-value-added but necessary work, and optimize worker safety. A company just beginning its Lean journey typically has no more than 65% of its frontline operators' time dedicated to value-added and non-value-added but necessary tasks. This provides an immediate opportunity to improve productivity by 35%. Further productivity improvements are achievable by eliminating or transferring non-valued-added activities performed by frontline operators to underutilized material handlers (typically, plant material handlers are 35% utilized). For example, usually in a plant, forklift trucks or powered pallet trucks have no load either going to or returning from the warehouse. This means at least 50% of their activity is waste. Standardized work and cell design along with other Lean practices will significantly improve value-added time percentages of all frontline team members.

When we think of standards and standardized work, we automatically think of frontline operations. In the LEOMS, all team members have well-defined standardized work, which is critical to accountability and achievement of results through waste elimination. Standard work is defined for every level of management, at least for the portion of their job that is an integral part of the LEOMS. Thus, managers must have structured time in their daily schedules for frontline time. Daily schedules must be arranged so managers review all operations over a defined period. The purpose of this is to:

1. Audit standardized work
2. Coach team members in the LEOMS, particularly areas most relevant to their workplace and continuous improvement
3. Follow up on previously identified deviations to ensure that corrective action is being completed
4. Identify the next level of system improvement

This is a critical management activity required for building and sustaining the LEOMS. A second standardized work activity for managers is leading policy deployment–Hoshin Kanri each year, ensuring that operations are completely aligned with annual business plan improvement targets. The third manager's standardized work activity is conducting regular Lean operational reviews. Regular reviews are required to ensure organizational accountability for achievement of operational plan targets and coaching team members. Coaching is the fourth management requirement and a daily responsibility to achieve success with Lean continuous improvement, as leading by example is the only path to changing a company's people management culture and practices.

Improvement Progress through Wave 4

Improvement progress in achieving annual improvement targets should start to appear after waves 3 and 4 on safety, purchased material inventories, productivity, and throughput, resulting in lower operational cost and purchased materials inventories (see Figure 6.39). Safety should improve as every work cell operation has been redesigned based on ergonomic and industrial engineering best practices, eliminating wasted motion and activities. Productivity gains mean some frontline team members can be redeployed to support continuously operating production through shift changes and by covering for vacation and sick days, lunch breaks, and training, and providing time for participation in group problem solving. This will result in improving OEE through more output per shift. This is an important step as team members see management actions confirm their words related to investing in frontline team members. It is not time to sit back and enjoy the improvements, but to use the organizational energy created from these successes to accelerate PDCA activities and training. Job security only exists in companies that have sustained market success. Having discussions with team members about their role in creating company success and their own job security as valued team members should be undertaken to reinforce their role as "co-owners" of the operation with a voice as collaborative partners in decisions that create sustainable success. With work cells optimized, linking the LEOMS together in wave 5 creates value stream synchronization to the customers' rhythm, contributing to improved service, operational cost, and inventory reductions. Completion of wave 4 (see Figure 6.40) created productivity, safety, and cost improvement in work cell performance, and while it is an important step, there is significant hard work remaining to achieve a mature LEOMS. Wave 5 kicks off with a focus on building out the system by connecting work cells into an integrated materials flow system.

Annual Supply Chain Operatons Strategic Plan Targets to Improve																	
Each target you have previously entered will appear below automatically. For each target, enter your starting point and plot the progression needed to meet the target on time. Fill in the actual progression over time. Current cell formatting will show																	
Project Status	Last Status Review Date:		Green	On-plan	Yellow	At risk	Red	Off-plan	Initiate		Continue					
Priority	Deliverables and measures	Starting point	Month	J	F	M	A	M	J	J	A	S	O	N	D		
1	Reduce RIR 2.1 to 1.7 LITR 1.75 to 1.63	2.1% R.I.R. 1.75% L.I.R.	Plan		On-plan	On-plan	On-plan	On-plan	On-plan	On-plan	On-plan	On-plan	On-plan				
			Actual	2.11 1.77	2.09 1.74	2.10 1.73	2.08 1.72	1.90 1.68	1.86 1.65	1.80 1.53	1.75 1.46	1.63 1.35	1.43 1.21				
2	Reduce factory cost by 1% point	68.0%	Plan		On-plan	On-plan	On-plan	On-plan	On-plan	On-plan	On-plan	On-plan	On-plan				
			Actual	68	68.2	68	67.9	67.8	67.5	67.6	67.4	67%	66.6				
4	Improve order release to shipment leadtime by 1 day	5 Days	Plan		At risk	On-plan	On-plan	On-plan	On-plan	On-plan	On-plan	On-plan	At risk				
			Actual	5	5.2	4.8	4.6	4.7	4.4	4.5	4.3	4.0	4.2				
7	Produce $ 30M of cash by reducing inventory	$0	Plan		On-plan	On-plan	On-plan	On-plan	On-plan	On-plan	On-plan	On-plan	On-plan				
			Actual	$0	$0	$0	$0	$0	$2.0M M	$4.0M M	$10M M	$15M M	$20M M				

Figure 6.39 MMC annual supply chain targets to improve.

For each target, enter your starting point and shade in the number of months to complete each wave. Monthly plot progress in achieving the milestones for each month. If progress has not achieved the milestones a corrective action plan is required to be implemented to get back on plan.

Priority	Deliverables and measures	Starting point		1	2	3	4	5	6	7	8	9	10	11	12	13	14	15	16	17	18	19	20	21	22	23	24
1	Initiate culture change plan implementation month 1		Plan																									
			Actual																									
2	Initiate wave 1 month 2		Plan																									
			Actual																									
3	Initiate wave 2 month 3		Plan																									
			Actual																									
4	Initiate wave 3 month 5		Plan																									
			Actual																									
5	Initiate wave 4 month 8		Plan																									
			Actual																									
6	Initiate wave 5 month 11		Plan																									
			Actual																									
7	Initiate wave 6 month 12		Plan																									
			Actual																									
8	Initiate wave 7 for annual strategic planning process		Plan																									
			Actual																									
9	Create sufficient improvement cost benefits to fund year 1 startup cost		Plan																									
			Actual																									

Figure 6.40   LEOMS dashboard.

Wave 5

Creating Synchronized Flow

Wave 5 (see Figure 6.41) connects the LEOMS together, creating a supply chain synchronized to the customer's demand rhythm. This results in an opportunity to synchronize resource consumption as material flows and work cells are integrated together creating a system with a zero-waste target state. While this may

Wave 5 system deployment ⑤

Value stream system deployment wave deliverables:
- Pacemaker scheduling methods implemented
- Plant level scheduling process – Heijunka
- Inventory supermarkets and Kanban system replenishment for in-process and finished goods inventory
- Integrated material delivery routes supporting plant value streams
- Standard work instructions indirect team member tasks
- Optimized and balanced work among cell and indirect team members
- Continue building Ninjutsu

Information required
- Plan for every part
- Takt time
- VSM current and future state
- Current and 3 year volumes (products defined at the level necessary to identify what will be required from the process)
- Annual and 3 year performance improvement targets for key business and operational metrics
- Capacity and supply chain strategy – define based on demand profiles, volume seasonality, service strategy
- Three year facility plan

Figure 6.41   LEOMS wave 5 system integration.

never be achieved, the mere fact that the LEOMS creates this potential means waste can be eliminated forever in pursuing system perfection. *Creating Level Pull*[11] by Art Smalley is an excellent guide to implementation of wave 5 as it covers the wave 5 deliverables.

Creating Level Pull across Value Stream Systems

The process starts with having the right information; in this case, it is PFEP, the Plan For Every Part database. For example, data from PFEP are used to determine inventory levels, storage container sizes, and the number of containers in the system for each product. In the case of MMC, pitch was calculated based on one pallet of finished product. This could be a full pallet of one product or a pallet quantity made up of two products, one with two levels of finished goods and another with three levels to create an "equivalent pallet."

Creating Finished Good Supermarkets

The process starts by using an ABC analysis of all finished good items to determine whether the plant supply chain will be make-to-order, make-to-stock, or a hybrid system. Major questions usually relate to "C" or low-volume items: should they be make-to-order items or remain as make-to-stock items? It is also important to think about the potential impact on customers as part of the final direction related to make-to-order or make-to-stock for "C" items. With continuous change going on in a plant that is implementing the LEOMS, a legitimate option is to maintain all items as stock items until the LEOMS is stable and has matured to a point where there is a comfort level in moving to a hybrid system without risking customer service failures. MMC will start with all items being stock items and accept higher "C item" inventory cost, and reconsider this decision after their entire system has achieved a few months of stability. Completing the first step of building a finished good supermarket will enable implementation of pacemaker and heijunka at assembly; again see pages 27 through 46 of the book *Creating Level Pull*.[11]

Regulating the System Rhythm—Pacemaker

During the process of creating MMC's future state VSM, assembly was defined as the pacemaker process with a Takt time of 1.512 seconds and a scheduling pitch of 1512 seconds or 25.2 minutes. Pitch would consist of either a single pallet, 100 cartons of one product, or a combination of 40 cartons of one product and 60 cartons of another product. See black hexagon 1 in Figure 6.42. Production kanban cards from shipping are picked up by production control each morning and placed in the heijunka box in priority order for assembly to fulfill. In some processes, it is necessary to produce products in a certain sequence and/or families of products need to be produced together. In these cases, a changeover wheel is

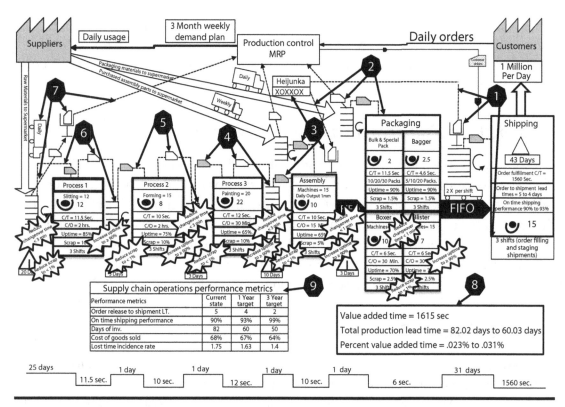

Figure 6.42 MMC generation 1 current state improvement projects.

employed by production control to manage sequencing of products to be produced and calculation of inventory required in the next downstream process's supermarket to ensure availability to sustain value stream production at Takt time pace.

In-Process Supermarkets

Upstream of the pacemaker operation, assembly in MMC's case, in-process supermarkets are implemented as a part of completing projects 3, 4, 5, and 6 (see Figure 6.42). Supermarkets are not a value-added process but often are required to support a value stream's smooth flow as single piece continuous flow is not possible when processes have differences in cycle times, changeover times, and up-time percentages. See pages 27 through 46 of the book *Creating Level Pull*[11] for instructions on creating in-process supermarkets.

Create Integrated Material Delivery Routes

Integrated material pickup and delivery routes typically involve multiple team members assigned to delivery and pickup routes at scheduled time intervals, assuring that materials are available at their points of use when needed. (See "Expanding the System", pages 73–88, of *Creating Level Pull*[11] for instructions on

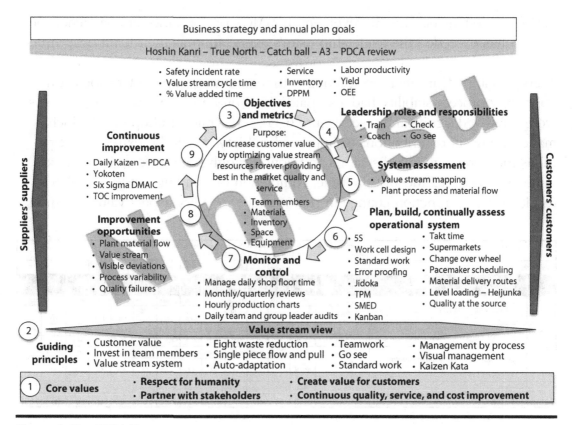

Figure 6.43   LEOMS.

building out level pull systems throughout a value stream.) Implementation of these routes is the final step in connecting an entire value stream system together operating at Takt time. MMC process engineers, frontline team members, along with team and group leaders focused all their time and effort during wave 5 on improving process stability to reduce scrap, increase yields and uptime percentages, and defer restructuring workplace layouts until the next cycle through the LEOMS's waves. With completion of wave 5, MMC's basic Lean enterprise value stream system is complete, but this journey is just beginning; look through the details shown in Figure 6.43 and see that all the LEOMS's operating elements are in place.

The hardest work is yet to be done: building an army of continuous improvement scientists that will result in creating Ninjutsu the "management majic". Wave 6 focuses on building the army of scientists who will make continuous improvement into a powerful competitive weapon.

MMC Generation 2 Current State Value Stream Map

At the end of wave 5, the generation 2 current value stream map is created as the basis for the second round of system improvements, and it reflects the Lean enterprise operational system's status and establishes new one and three year performance targets. The improvement in MMC's performance at completion of wave 5 has exceeded the leadership team's expectation as their team beat three of the four year 1 plan targets to improve (see Figure 6.44). MMC process engineers', frontline team members', along with team and group leaders' focus on improving process stability had also paid off as process uptimes and scrap rates were improved significantly (see Figure 6.45); scrap rates and uptimes, which made significant contributions to improved stability, would be extremely helpful as they optimized workstations and flow through each process in the next improvement cycle. Most of the time, completion of building out the LEOMS results in a step function change in performance in every aspect value stream

	Each target you have previously entered will appear below automatically. For each target, enter your starting point and plot the progression needed to meet the target on time. Fill in the actual progression over time. Current cell formatting will show															
Project Status	Last Status Review Date:			Green	On-plan	Yellow	At risk	Red	Off-plan	Initiate		Continue			
Priority	Deliverables and measures	Starting point	Month	J	F	M	A	M	J	J	A	S	O	N	D
1	Reduce RIR 2.1 to 1.7 LITR 1.75 to 1.63	2.1% R.I.R. 1.75% L.I.R.	Plan		On-plan	On-plan	On-plan	On-plan	On-plan	On-plan	On-plan	On-plan	On-plan	On-plan	
			Actual	2.11 1.77	2.09 1.74	2.10 1.73	2.08 1.72	1.90 1.68	1.86 1.65	1.80 1.53	1.75 1.46	1.63 1.35	1.43 1.21	1.30 .84	
2	Reduce factory cost by 1% point	68.0%	Plan		On-plan	On-plan	On-plan	On-plan	On-plan	On-plan	On-plan	On-plan	On-plan	On-plan	
			Actual	68	68.2	68	67.9	67.8	67.5	67.6	67.4	67%	66.6	66.3	
4	Improve order release to shipment leadtime by 1 day	5 Days	Plan		At risk	On-plan	On-plan	On-plan	On-plan	On-plan	On-plan	On-plan	At risk	At risk	
			Actual	5	5.2	4.8	4.6	4.7	4.4	4.5	4.3	4.0	4.2	4.1	
7	Produce $ 30M of cash by reducing inventory	$0	Plan		On-plan	On-plan	On-plan	On-plan	On-plan	On-plan	On-plan	On-plan	On-plan	Off-plan	
			Actual	$0	$0	$0	$0	$0	$2.0MM	$4.0MM	$10MM	$15MM	$20MM	$22MM	

Figure 6.44 Annual supply chain targets to improve.

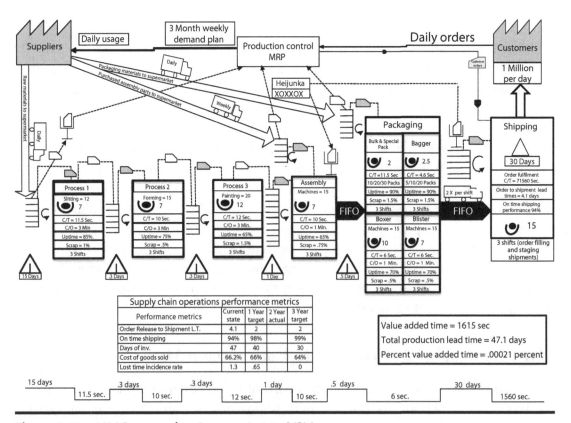

Figure 6.45 MMC generation 2 current state VSM.

performance. The organization will be energized by what they have accomplished, so this moment must be seized by leadership to reflect on improvements achieved, praise the organization for their hard work and results, and challenge them that "we" can do more to overcome the hard challenges that are yet to be faced by:

- Building organizational daily discipline to sustain the gains
- Sustaining the drive for perfection every day
- Creating the army of continuous improvement scientists
- Being conscious that our daily behavior is consistent with the people management culture required to successfully develop and train all team members, so Ninjutsu appears every day across all operations

Wave 6

Continuous Improvement

With wave 5 complete the process moves on to wave 6 (see Figure 6.46). LEOMS wave 6 continuous improvement (Figure 6.47) initiates system continuous improvement and maturation of the process of grafting continuous improvement genes into the organizational culture. Wave 6 deliverables are mainly focused on maturing daily practices that include "continuous improvement routines" through coaching and training of all organizational levels. It is advisable to create a behavior and skills matrix showing all required LEOMS practices that must be mastered by each organizational level and apply it as a specific position related professional development assessment to assure that needed skills are being developed. The expectation of all team members is to develop their knowledge and

For each target, enter your starting point and shade in the number of months to complete each wave. Monthly plot progress in achieving the milestones for each month. If progress has not achieved the milestones a corrective action plan is required to be implemented to get back on plan.

Priority	Deliverables and measures	Starting point	Plan/Actual	1	2	3	4	5	6	7	8	9	10	11	12	13	14	15	16	17	19	20	21	22	23	24
1	Initiate culture change plan implmentation month 1		Plan																								
			Actual																								
2	Initiate wave 1 month 2		Plan																								
			Actual																								
3	Initiate wave 2 month 3		Plan																								
			Actual																								
4	Initiate wave 3 month 5		Plan																								
			Actual																								
5	Initiate wave 4 month 8		Plan																								
			Actual																								
6	Initiate wave 5 month 11		Plan																								
			Actual																								
7	Initiate wave 6 month 12		Plan																								
			Actual																								
8	Initiate wave 7 for annual strategic planning process		Plan																								
			Actual																								
9	Create sufficient improvement cost benefits to fund year 1 startup cost		Plan																								
			Actual																								

Figure 6.46   LEOMS implementation dashboard.

Continuous improvement deliverables:
- Structured quarterly Lean review process including 90 day improvement plan
- Ongoing coaching of champions and plant managers
- Increased depth of Lean understanding
- Follow up training as needed
- Scheduled shop floor coaching by all levels of leadership
- Lead and encourage sharing learnings from PDCA across the organization
- A3 corrective action follow up for business and operational leadership
- Implement low cost 5 year facility plan changes
- Repeat waves 2 through 6 through rigorous daily application of the improvement routines and coaching, forever
- Leaders demonstrate they are living the Lean culture

Information required
- Structured plant manager 90 day progress reviews
- Business and total operations Bowler 90 day progress reviews
- LEOMS implementation red-green charts on plant progress in achieving yr. 1 metric targets

Figure 6.47   LEOMS wave 6 continuous improvement.

practice their work skills and coaching skills by investing time in coaching team members. This expectation needs to be part of annual performance reviews, and Lean coach development assessments as presented earlier in the chapter should be utilized for this purpose. Review Figure 6.10 Lean Coach Development Assessment have every person in a leadership position complete a self assessment before starting wave 6.

Organizational PDCA Responsibilities

In the book *Toyota Culture*[12] (pp. 178–179), the authors quote a study conducted by Professor Koichi Shimizu of Okayama University that 90% of realized improvements are carried out by frontline team and group leaders and assigned manufacturing engineers, and 10% comes from frontline team members. Frontline team members are focused on their work assignment responsibilities and executing them to maintain output at the defined Takt time. Because of this focus, most of their improvement ideas are related to their own cell work responsibilities, which limits the financial impact and significance of their improvements. But this does not diminish the importance and value of their contribution for two reasons:

1. Being responsible for continuous improvement utilizes the full capability of frontline team members and making them full and equal team members.
2. As the authors go on to explain the other main purposes of frontline team members, problem-solving responsibility is knowledge and skill development to identify prospective group and team leaders.

Development of frontline team members' knowledge and skills also means that they are consciously improving their "eyes to see" process variability and

Job level	Process problem solving	PDCA focus	Business planning
Department head	Pattern problem solving	System Kaizen	Company and department priorities
Manager and group leaders	Process performance problems	Process performance management	Identify section priorities and improvement
Team member	The work area process problems	Process standards and targets	Daily area of responsibility work assignment

Figure 6.48 PDCA organizational level focus. (From Jeffery Liker and Michael Hoseus *Toyota Culture*, **p. 179, McGraw-Hill, 2008.)**

We don't think our way to a new way of acting but act our way to a new way of thinking

From: Implement new principles, techniques, practices

To: Implement new principles, techniques and practices by building consistent organizational behavior patterns resulting in an army of continuous improvement scientist and new company culture attributes

Figure 6.49 A different challenge—We act our way to a new way of thinking. (Adapted from Mike Rother, *Toyota Kata*, **pp. 232. McGraw-Hill, 2010.)**

defects in their daily work assignments, making them better contributors to identifying larger system failures for team leaders, group leaders, and manufacturing engineers to solve (see Figure 6.48). In *Toyota Kata*,[5] Mike Rother and Michael Hoseus provide readers with examples of the "routines" needed to build an army of continuous improvement scientists; these routines reshape company culture, inculcating the principle of managing by process and its accompanying routine of daily application of PDCA. A core principle of organizational change is illustrated in Figure 6.49; organizations do not think their way to a new way of acting, but act their way to a new way of thinking, reinforcing the importance of "kata" in building an army of problem-solving scientists.

Continuous Improvement Cycle

This continuous improvement cycle (Figure 6.50) is driven by daily routines of identifying waste and applying PDCA to resolve it. Over time, this means going deeper into the value stream system to standardize, level load, stabilize, and improve flow to a new level. Stability is equal to comfort, and it is not productive to allow organizations to get accustomed to comfort; so planned "destabilization" is employed to force another cycle of improvement to bring processes back to stability at a higher level of performance. Every process should have constantly

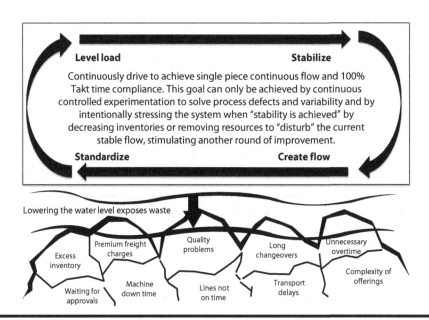

Figure 6.50   Lean system continuous improvement cycle.

revised target conditions and performance identified to challenge the organization to constantly improve process performance to new levels. Developing "problem consciousness" (see Figure 6.51[13]) as an everyday habit is critical to sustaining continuous improvement and combats the natural tendency we all have of just walking by problems because they have always existed. Developing critical eyes to see requires daily discipline and practice over a long time period to overcome our old paradigms and habits. The target conditions may involve less inventory, reducing a machine cycle time or removing an operator from a cell to achieve target results. Application of target conditions will provoke application of PDCA (see Figure 6.52) to development solutions, bringing processes back to stability under the new target conditions. This involves taking some level of risk, which potentially disrupts short-term product flow as faster cycles increase flow and raise a new series of barriers that must be addressed, thus driving requirements for continual improvement. A fundamental Lean implementation rule is "always protect the customer." This means when risk exists, sufficient project resources are applied to ensure that any problems can be dealt with immediately through application of temporary countermeasures until a permanent solution is defined

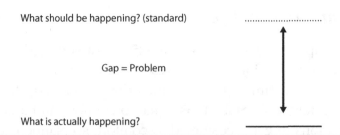

Figure 6.51   Problem consciousness. (From Pascal Dennis, *Getting the Right Things Done*, Cambridge, 2006.)

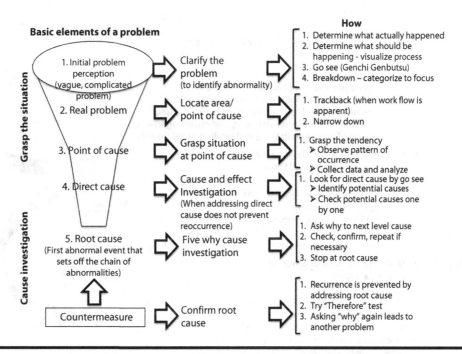

Figure 6.52   Practical problem solving.

and implemented. This value stream continuous improvement cycle is repeated by going deeper into the value stream system and/or operations to implement new target conditions to standardize, level load, stabilize, and improve flow to a new performance level by applying PDCA.

This drive to new levels of performance would typically be undertaken by the frontline leadership and manufacturing engineers in collaboration with the frontline team members. Participation by frontline team members provides a very valuable development experience giving them confidence to tackle problems that arise day to day that contribute to their professional development. The ultimate goal is to achieve kaizen mastery as shown in Figure 6.53. Kaizen mastery is achieved by practicing kaizen daily resulting in experiences that are invaluable as team members experience being valued not just from hearing words from executives about what they are contributing.

CONTINUOUS IMPROVEMENT AND ADAPTATION

What I mean when I say continuous improvement and adaptation: the ability to move toward a new desired state through an unclear and unpredictable territory by being sensitive to and responding to actual conditions on the ground.

Mike Rother
Toyota Kata, McGraw-Hill, pp. 8–9, 2010

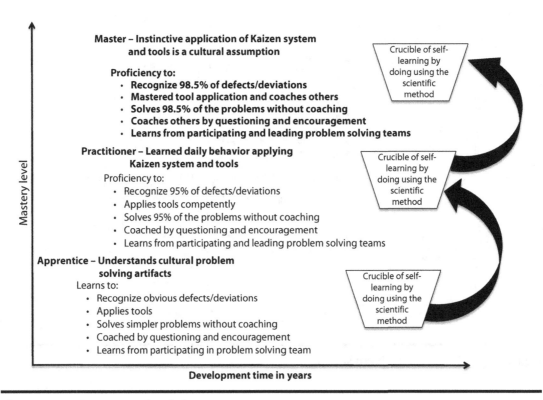

Figure 6.53   Team member kaizen mastery.

It is these valued experiences of team members that develop their perspective on their role as valued team members who have a say in their day-to-day job. These developmental experiences are the source of Ninjutsu (Figure 6.54) as frontline team members gain greater skills and feel a part of a bigger team. Employee relations principles underpinning the LEOMS's Ninjutsu are proven solid principles that should be adopted by every company. With the LEOMS first pass through implementation complete, MMC's performance improved significantly (see Figure 6.55), as every plant frontline operational process has been redesigned to eliminate waste, and three of four annual targets to improve were achieved; the fourth was just 1/10 of a point above target. The best is yet to come for MMC as they have shown sufficient commitment and persistence to develop

---

**LEAN CULTURE INVESTS IN PEOPLE TO GET
A RETURN FROM THEIR APPLIED KNOWLEDGE**

One of the biggest elements of cultural change is not just paying hourly workers for their hands but their minds as well. Getting rid of the eighth type of waste "Lack of return on knowledge". Then training the entire workforce on Lean principles and practical problem solving. Which is what NUMMI did!

Jerome Hamilton

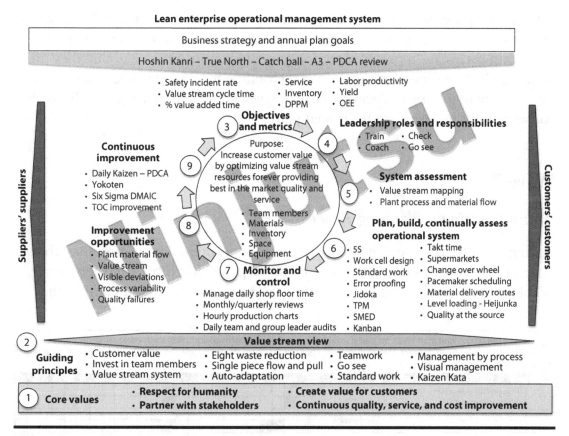

Figure 6.54 Ninjutsu.

Project status	Last status review date:			Green	On-plan	Yellow	At risk	Red	Off-plan	Initiate		Continue				
Priority	Deliverables and measures	Starting point	Month	J	F	M	A	M	J	J	A	S	O	N	D
1	Reduce RIR 2.1 to 1.7 LITR 1.75 to 1.63	2.1% R.I.R. 1.75% L.I.R.	Plan		On-plan	On-plan	On-plan	On-plan	On-plan	On-plan	On-plan	On-plan	On-plan	On-plan	On-plan
			Actual	2.11 1.77	2.09 1.74	2.10 1.73	2.08 1.72	1.90 1.68	1.86 1.65	1.80 1.53	1.75 1.46	1.63 1.35	1.43 1.21	1.30 .84	1.10 .65
2	Reduce factory cost by 1% point	68.0%	Plan		On-plan	On-plan	On-plan	On-plan	On-plan	On-plan	On-plan	On-plan	On-plan	On-plan	On-plan
			Actual	68	68.2	68	67.9	67.8	67.5	67.6	67.4	67%	66.6	66.3	66.2
4	Improve order release to shipment lead time by 1 day	5 days	Plan		At risk	On-plan	On-plan	On-plan	On-plan	On-plan	On-plan	On-plan	At risk	At risk	At risk
			Actual	5	5.2	4.8	4.6	4.7	4.4	4.5	4.3	4.0	4.2	4.1	4.1
7	Produce $ 30MM of cash by reducing inventory	$0	Plan		On-plan	On-plan	On-plan	On-plan	On-plan	On-plan	On-plan	On-plan	At risk	Off-plan	Off-plan
			Actual	$0	$0	$0	$0	$0	$2.0M M	$4.0M M	$10M M	$15M M	$20M M	$22M M	$25M M

Annual supply chain operations strategic plan targets to improve
Each target you have previously entered will appear below automatically. For each target, enter your starting point and plot the progression needed to meet the target on time. Fill in the actual progression over time. Current cell formatting will show

Figure 6.55 Wave 6 targets to improve status.

an army of scientists making daily small process improvements that aggregate to provide their company continuous improvement in competitiveness and financial resources to reinvest in growing the business. This is also a good time to review opportunities to invest in facility changes that will better support long-term value stream operational cost and cash improvement results. It is now time to start another cycle of system improvement starting with wave 2. While frontline continuous improvement should be ongoing, formally going through the LEOMS's wave cycle as the primary approach to raising the bar on system design and performance through building on experiences from the first pass prepares team members to be more active participants in all aspects of continuously improving

MMC's enterprise operational management system. Management involvement and engagement in Hoshin should continue to increase and mature as more implementation cycles are repeated, and operational team members feel that they are an integral part of the planning process. Frontline team member learnings can be great contributions to deciding what "bigger improvement projects" should be considered for inclusion in the annual plan.

Wave 7

Annual Strategic Planning Process

Having completed wave 6, it is time to move on to preparing to repeat another cycle of the six waves (see Figure 6.56). This step is initiated by completing wave 7 (see Figure 6.57), an assessment of culture change progress and a plan for the

For each target, enter your starting point and shade in the number of months to complete each wave. Monthly plot progress in achieving the milestones for each month. If progress has not achieved the milestones a corrective action plan is required to be implemented to get back on plan.				

Priority	Deliverables and measures	Focus		1 2 3 4 5 6 7 8 9 10 11 12 13 14 15 16 17 19 20 21 22 23 24
1	Initiate culture change plan implementation month 1	Reinforce culture	Plan / Actual	
2	Initiate wave 1 month 2	Stretch plan	Plan / Actual	
3	Initiate wave 2 month 3	Accelerate improvement	Plan / Actual	
4	Initiate wave 3 month 5	Safety first	Plan / Actual	
5	Initiate wave 4 month 8	Accelerate improvement	Plan / Actual	
6	Initiate wave 5 month 11	Accelerate improvement	Plan / Actual	
7	Initiate wave 6 month 12	Make it the norm	Plan / Actual	
8	Initiate wave 7 for annual strategic planning process	Make it the norm	Plan / Actual	
9	Deliver cost and cash to cover inflation cost and a 20% R and D budget increase	Be the market leader	Plan / Actual	

Figure 6.56   LEOMS implementation dashboard.

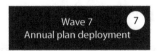

Wave 7
Annual plan deployment   7

Lean strategy to operations alignment deliverables:
- A more effective annual operational planning process
- All division resources fully aligned and focused on critical programs and strategic plan goals
- Methods and tools to implement the process
- Management implementation training
- Coaching during one complete annual planning and review cycle

Information required:
- Strategic business plan
- Annual business plan objectives and key programs
- Critical assumptions required for annual plan success – what conditions must exist?

Figure 6.57   LEOMS wave 7.

coming year that is developed by conducting a deep assessment of organizational progress and identification of what actions need to be taken to reinforce first year progress and set operational objectives for the coming year, and what improvements will be made to the continuous improvement process. The progress assessment should be developed using the culture change strategic actions to be taken developed in Chapter 4 (Figures 4.23 and 4.24). It is reasonable to continue including culture change progress assessments until at least two consecutive years of assessments verify that the changes have truly become a part of the organization's behavioral fabric. It is also useful to republish culture change expectations to every organizational level and have small group sessions to discuss progress, review changes made, and get input from team members on their biggest challenges. It is also suggested to get input on what top leadership can do to assist them in making change a part of daily organizational behavior. Starting in year 2 of implementation, 60% of projected financial benefits from continuous system and operations, should be included in the annual financial plan. Why 60%? Inflationary cost must be covered before net P&L and capital employed benefits are gained. Secondly, there is always some "leakage" in the projected actual savings. This also increases commitment from executives and their leadership teams as their performance appraisal ratings reflect their contribution to culture change, increased value to customers, and continuous cost reduction goal achievement.

Build Expected Benefits into Annual Business Planning

MMC's organization developed their confidence in the LEOMS during their first year and looked forward to year 2 with enthusiasm to "raise the bar" on improvement results. Their commitment to increasing improvement levels was reflected in their year 2 strategic plan targets to improve (see Figure 6.58). Safety would continue to be improved by achieving a 10% reduction in both R.I.R. and L.T.I.R, another 0.5 percentage point reduction in cost of goods sold, achieving 2-day order

---

**BUILD 60% OF THE EXPECTED GAINS
INTO THE YEAR TWO FORECAST**

When Jim McNerney brought Six Sigma to 3M, he demonstrated his personal commitment to the program and made the investments necessary to get it up and running. But, his second big move got the attention of every leader in the company. Starting the second fiscal year after the program started, 60% of the projected savings was included in the company's official financial forecast. This elevated the commitment by all leaders to deliver results. 3M CEO Jim McNerney ups the ante to get 3M leaders' skin in the game.

Paul Husby

Annual supply chain operatons strategic plan targets to improve																			
Each target you have previously entered will appear below automatically. For each target, enter your starting point and plot the progression needed to meet the target on time. Fill in the actual progression over time. Current cell formatting will show																			
Project status	Last status review date:			Green	On-plan	Yellow	At risk	Red	Off-plan	Initiate		Continue							
Priority	Deliverables and measures	Starting point	Month	J	F	M	A	M	J	J	A	S	O	N	D				
1	Reduce RIR from 1.10% to .90% Reduce L.I.R. from .65% to .45%	1.10% R.I.R. .65% L.I.R.	Plan																
			Actual																
2	Reduce cost of goods sold from 66.2% to 65.7%	66.2%	Plan																
			Actual																
4	Reduce order release to shipment L.T. from 4.1 days to 2 days	4.1 days	Plan																
			Actual																
7	Increase net cash flow from $250K to $1.0 MM	$250	Plan																
			Actual																

Figure 6.58 MMC strategic plan year 2 targets.

fulfillment lead time, and generating an additional $10 million in cash by reducing inventory. MMC is on their way to reestablishing their market leadership position.

Summary

Implementation of the LEOMS is a huge challenge for any organization, requiring top leadership's consensus and full commitment, as any "going-through-the-motions leadership" will damage progress; so it is imperative that top leadership performance appraisals and incentive compensation plans are linked to LEOMS implementation success. Chapter 5 along with this chapter presented implementation plan content with sufficient detail and material references that provide readers with information and references to develop themselves into high level LEOMS practitioners. This is not to say assistance and training from a qualified Lean consulting organization is not important or unnecessary, but that becoming self-sufficient as an organization requires developing a cadre of highly qualified LEOMS practitioners who are responsible for training, development, and assessments of both LEOMS progress along with a LEOMS development and certification program for all organizational levels. It is easy to become attached to the LEOMS practices and tools as a company sees great results achieved by applying them, but it is important to keep in mind that without nurturing the LEOMS compatible culture, none of that will be sustainable. Top executives should delegate the LEOMS seven wave deployment to their next level team members, but take personal ownership for making necessary culture changes within the organization and being an example of behaviors and management style exemplifying in their everyday interactions with team members.

References

1. Supply Chain Council, SCOR 9.0, 2009.
2. Mike Rother and John Shook, *Learning to See*, Cambridge, MA: Lean Enterprise Institute, 2003.

3. Shigeo Shingo, *A Revolution in Manufacturing: The SMED System*, New York: Productivity Press, 1985.
4. Rick Harris and Earl Wilson, *Making Materials Flow*, Cambridge, MA: Lean Enterprise Institute, 2003.
5. Mike Rother, *Toyota Kata*, Burr Ridge: McGraw-Hill, 2010.
6. Durward K. Sobek and Art Smalley, *Understanding A3 Thinking*, Boca Raton, FL: CRC Press, 2008.
7. Productivity Press Development Team, *5S for Operators*, New York: Productivity Press, 1996.
8. Rick Harris and Mike Rother, *Creating Continuous Flow*, Cambridge, MA: Lean Enterprise Institute, 2001.
9. Steven Spear, *The High Velocity Edge*, Burr Ridge: McGraw-Hill, 2009.
10. Taiichi Ohno, *Toyota Production System*, New York: Productivity Press, 1988.
11. Art Smalley, *Creating Level Pull*, Cambridge, MA: Lean Enterprise Institute, 2004.
12. Jeffery Liker and Michael Hoseus, *Toyota Culture*, Burr Ridge: McGraw-Hill, 2008.
13. Pascal Dennis, *Getting the Right Things Done*, Cambridge, MA: Lean Enterprise Institute, 2006.

Chapter 7

Applying the LEOMS to Enterprise Supply Chains

What Will I Learn?

How to apply LEOMS principles and tools to assess, design, and improve supply chains.

In Chapter 6, the LEOMS was applied to factory operations and processes, but there is a broader dimension to supply chains, including a network of suppliers and customers spread across a country and or the globe, supported by processes and information technology systems. The LEOMS must be applied to entire supply chains to continuously improve their design, operation, and performance, ensuring continued delivery of unmatched value to customers. In reality, typical supply chain designs remain unchanged for many years even as they evolve with more products, customers, and suppliers. Supply chain requirements are constantly changing; suppliers and supplier performance, customers and their expectations, product designs, production processes, and business rules continuously evolve. These changes make their way into supply chains, and frequently, their negative impact on the current design goes unrecognized. They have the effect of obsoleting original supply chain design assumptions. To prevent effects of supply chain evolution from negatively affecting supply chain performance requires:

1. A robust high level supply chain dashboard to maintain clarity and focus with the few critical targets supported by multilevel metrics to facilitate rapid determination of specific processes and tasks affecting top-level critical few performance metrics.
2. Regular supply chain assessments, as very few people in the organization have their eyes on their entire supply chain because most people work in their own functional corner of the business.

3. Establishing one position with responsibility for the end-to-end supply chain at the highest organizational level overseeing end-to-end performance assessments, and to communicate the whole supply chain system view organization wide so everyone involved can see how their set of assigned tasks relate to overall supply chain performance.

4. Assigning dedicated resources to make regular assessments, model supply chains, stimulate improvement ideas, evaluate proposed changes, and recommend changes including redesign to continuously improve supply chain performance.

If current models are working, the tendency is to gravitate toward staying with them unless someone intervenes or a crisis occurs. Unfortunately, this approach ensures that new supply chains will follow the current designs with all of their problems, issues, and shortcomings. It takes vision, leadership, and resources to continuously rethink, improve, and redesign existing supply chains, their processes, and their supporting IT systems.

What is a supply chain? A supply chain is defined by a product family or families supplied to a group of customers with similar product and supply chain requirements; three well-recognized customer groups are shown in Figure 7.1. There are other customer groups such as government, educational institutions, financial institutions, military, etc., and it is often necessary to define subgroups of customers because of significant differences in their buying decision criteria and supply chain requirements. For example, some customers in the same market may place cost as their primary decision factor, while others choose based on service, product design, or product quality. What is important in segmenting customers for the purpose of defining supply chains is focusing on significant buying decision factor differences between groups of customers within markets. These significant differences need to be integrated into supply chain designs and operations to enable consistent customer service satisfaction.

Customer groups product families ➡ ⬇	End users - consumers	OEM	Channel partners
Product family 1	X	X	
Product family 2		X	X

Figure 7.1 What is a supply chain?

SCOR Supply Chain Design, Assessment, and Improvement Processes

SCOR, the supply chain operations reference model, is a product of the Supply Chain Council,[1] a nonprofit global corporation with global membership of companies and professionals who are interested in advancing and applying state-of-the-art supply chain management system practices. The Supply Chain Council provides training and certification to prepare professionals to understand and apply SCOR. The scope of SCOR includes all aspects of supply chain design and operations (see Figure 7.2). At its highest level, level 1, are major processes that exist in every manufactured product supply chain: plan, source, make, and deliver. SCOR provides practitioners a framework to ensure that business strategy drives supply chain configuration design, processes, metrics, best practices, and defines information technology functionality, creating a unified structure supporting supply chain design, operation, and improvement. SCOR's content is shown in Figure 7.3.

Supply chain configuration:

Level 1: Basis of competition translates the business strategy into a supply chain process type, make-to-stock, make-to-order, or engineer-to-order with prioritized attributes to support the chosen strategy.

Level 2: Supply chain configuration defines and configures the high-level supply chain processes to achieve the design attribute performance levels defined in levels 1 and 2.

Level 3: Defines target levels of performance, processes, and practices of daily operations and system configuration elements.

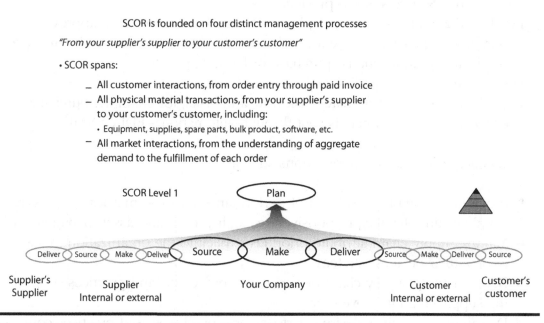

Figure 7.2   SCOR supply chain operations reference model.

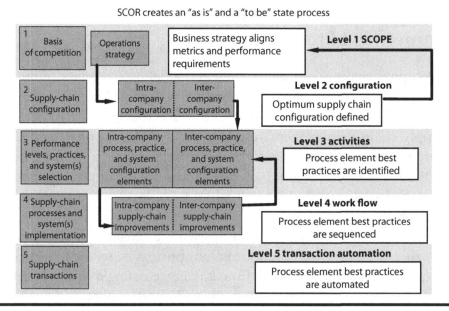

Figure 7.3 SCOR improvement process.

Level 4: Sequences the process practice tasks and system steps required to complete the process activities defined in level 3.

Level 5: Defines IT automation requirements to link and automate transactions in support of sequenced tasks defined in level 4.

SCOR structure example:

Level 1: S = Source.

Level 2: S1 = Source stocked product.

Level 3: S1.2 = Tasks—For example, create purchase requisition, approve purchase requisition, create a purchase order, issue purchase order, receive product, update inventory, purchase order systems, etc.

Level 4: Sequences steps in completing each task.

Level 5: Automation: Automates task flow and tasks; for example, purchase order release to suppliers is sent through automation—EDI or XML.

SCOR supports practical applications such as

- Providing a comprehensive set of performance attribute metrics to measure supply chain reliability, responsiveness, agility, cost, and asset management
- Making supply chain attribute priorities explicit to the organization in support of their business strategy
- Benchmarking supply chain performance, processes, and practices supporting supply chain improvement
- Defining a company's desired performance, processes, and practices compared to the information technology systems implied processes in order to

determine if the system supports desired processes and practices and/or must be modified to support them

■ Providing a predefined root cause analysis path to assist in defining specific processes and practices that are the source(s) of supply chain failures and/or key to achieving a desired performance improvement target

SCOR Level 1

What are supply chain performance attributes and their related metrics? Supply chain attributes are shown in the first column of Figure 7.4, followed by the strategy they support and associated metric(s) in column 3. The power of applying the LEOMS to broader supply chains is its capability to optimize all attributes to their maximum potential with the only restriction being design characteristics that are impractical to change. The LEOMS rewrites our historical "either or thinking" and replaces it with "both and thinking," and SCOR equips practitioners with tools to apply LEOMS thinking, tools, and practices to entire supply chain designs and operations.

SCOR Levels 2 and 3

Level 2 provides a configuration toolkit (see Figure 7.5) to define supply chain process configurations based on the supply chain type: make-to-order, make-to-stock, or engineer-to-order. All companies have one of these three supply chain types, or a hybrid of two, or on rare occasions all three types. Automotive manufacturers, for example, operate a hybrid make-to-stock and make-to-order supply chain. All manufacturing company supply chains have four big high-level

Question: What is/are the most important attribute/s to achieve your Lean supply chain strategy?

	Attribute	Strategy	Overall supply chain metric
External	Reliability	Consistently getting the orders right, product meets quality requirement	Perfect order fulfillment
	Responsiveness	Consistent speed of providing products/services to customers	Order fulfillment cycle time
	Agility	Time to respond to significant external market changes	Supply chain flexibility Supply chain adaptability
Internal	Cost	Cost associated with managing and operating the supply chain	Cost of goods sold Total supply chain management cost
	Assets	Effectiveness in managing the supply chain's assets in support of fulfillment	Cash to cash cycle time Return on supply chain fixed assets

Figure 7.4   Supply chain performance attributes.

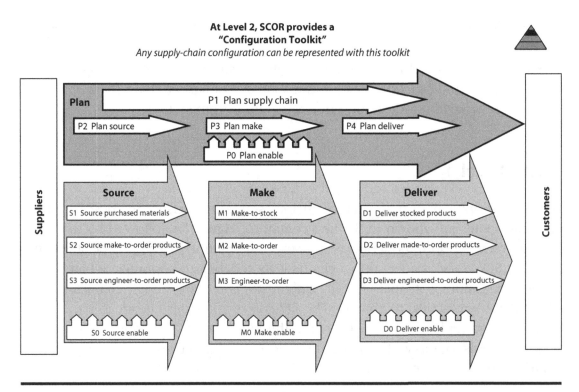

Figure 7.5   SCOR level 2 configuration tool kit.

processes: plan, source, make, and deliver. Specific plan, source, make, and deliver processes depend on their supply chain model, make-to-stock, make-to-order, or engineer-to-order. Level 3 defines the level 2 supply chain processes' detailed process elements (see Figure 7.6), including definition of process flow, inputs, outputs, metrics, practices, and enabling technology tools.

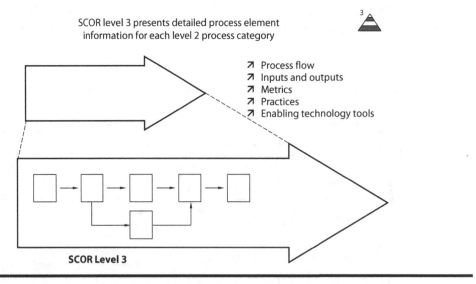

Figure 7.6   SCOR level 3 supply chain detailed process elements.

SCOR Level 4 Work Flow and Level 5 Automation

SCOR levels 4 and 5 sequence and automate process element practices defined in level 3 and are part of the requirements for defining and implementing effective processes. But because they are, by definition, "unique" to a particular company and supply chain situation, it is not possible to standardize all possible options; therefore, this is left up to organizations who are reengineering or designing their supply chain to define and document specific process work flow and its enabling automation. Level 4 activities are critical to designing or improving a supply chain as defining work flows along with their process activities from level 4 are inputs to configuring or modifying level 5 IT system automation to enable processes. This is one of the greatest causes of IT system implementation cost overruns, catastrophic startup failures, and disappointing operational results from IT investments. It is better to face this problem by working with IT suppliers to ensure that their system task-enabling details will operate in a manner that truly enables the desired level 3 process flow and tasks along with supporting required level 4 processes that support effective and efficient level 3 processes. Never make assumptions about how IT systems handle details of task activities, sequence, or flow as the risks are great that implementation failure and/or costly "system modifications" will be highly likely. Enabling technology is also required for the 10 enable elements shown in Figure 7.7. The first three elements relate to supply chain business rules, performance assessment, and data management; elements 4 through 10 enable processes in support of other company business processes, assuring that supply chain performance and priorities effectively deliver a company's business strategy, goals, and values.

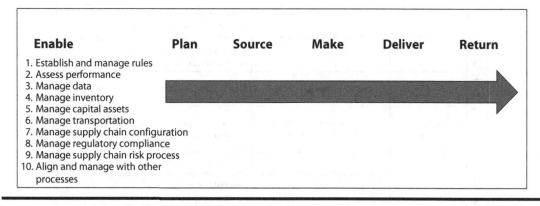

Figure 7.7   SCOR enable supply chain processes.

SCOR Metrics

SCOR focuses supply chain performance on five attributes, divided between internal and external attributes (see Figure 7.8):

Externally focused attributes:
- – Reliability
- – Responsiveness
- – Agility/flexibility

Internally focused attributes:
- – Cost
- – Assets

SCOR has three levels of metrics linked to internal and external supply chain performance attributes.

Level 1: Strategic metrics—the business key performance indicators
- – Measures overall supply chain performance, supply chain health
- – Establishes scope and objectives for a supply chain, project, or organization
- – Translates business problems or strategies into something measureable
- – Sets organizational priorities

Level 2: Performance metrics
- – Measure a part of a supply chain and/or strategic metric
- – Provide a first level of direction as to where improvement opportunities exist

Level 3: Diagnostic metrics: All metrics beyond level 1 and 2 metrics

Question: What are the operational metrics in support of achieving the Level 1 attribute goals ?

	Attribute	Strategic metrics	Supply chain operational metrics
External	Reliability	Perfect order fulfillment	Percent of order delivered on-time Percent orders delivered in full Percent orders delivered in perfect condition Percent orders received damage free Percent orders defect free
External	Responsiveness	Order fulfillment cycle time	Source, make, deliver product cycle times
External	Agility/ flexibility	Total production lead time percent 24/7 capacity utilization	Supply chain upside and downside adaptability Upside source, make and deliver flexibility
Internal	Cost	Total supply chain management cost	Cost of goods sold Cost to plan, source, make and deliver Direct labor cost Direct material cost
Internal	Assets	Cash to cash cycle time	Return on supply chain fixed assets Return on working capital

Figure 7.8 Supply chain level 1 and 2 metrics.

SCOR only specifies metrics for levels 1 and 2. Level 3 metrics serve as further diagnostic tools to assist in identification of specific processes and activities that are opportunities to improve performance. A useful guideline when identifying metrics is to set aside assumptions or current pressing problems, and seek to define the best metric possible, ones that best measure customer expectations precisely. Once these ideal metrics are defined, data collection methods must be considered. Measurement constraints are likely to appear, and it might be necessary to compromise. For example, "on-time delivery" may not be used unless actual delivery times are available for all shipments, so initially it may be necessary to use "on-time shipping". In addition to SCOR level 2 performance metrics, diagnostic metrics are defined at level 3 (see Figures 7.9 and 7.10). These metrics assist companies in identifying specific processes that are affecting supply chain performance. In effect, levels 1, 2, and 3 metrics are a predefined root cause analysis pathway, facilitating quick identification of specific processes and/ or practices that are likely source(s) of process failures and/or improvement opportunities.

SCOR provides a high-level project roadmap (see Figure 7.11) to assist in project development and execution. Because of SCOR's depth and detail, projects can appear to be overwhelming, so it is important to respect an organization's resource constraints by establishing a project scope and completion time frame that is challenging but does not overwhelm and discourage the organization. Projects frequently require IT system modification to enable chosen processes,

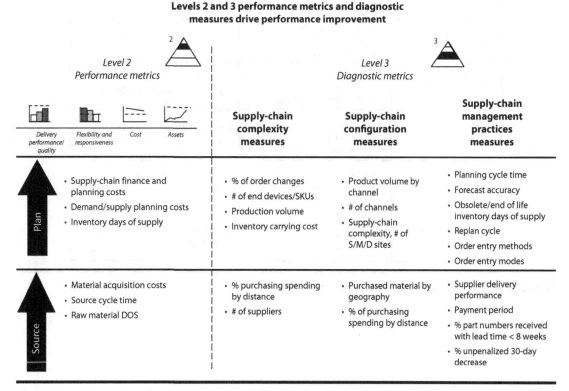

Figure 7.9   SCOR level 2 and 3 performance metrics: Plan and source.

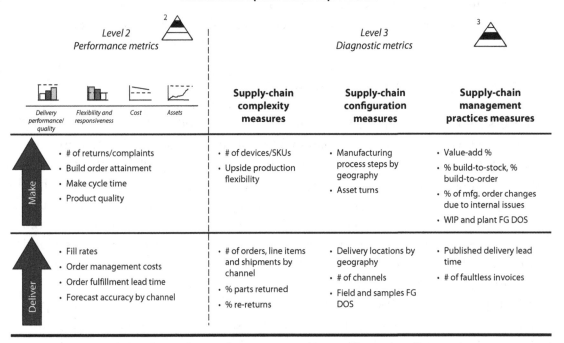

Figure 7.10   SCOR level 2 and 3 performance metrics: Make and Deliver.

SCOR project roadmap

Phase	Name	Deliverable	Defines
	Organize	• Organizational support	Who is the sponsor?
I	Discover	• Supply chain definition • Supply chain priorities • Project charter	What is the project scope?
II	Analyze	• Scorecard • Benchmark • Competitive requirements	What are the strategic requirements of the supply chain?
III	Material	• Geo map • Thread diagram • Disconnect analysis	Initial analysis: where are the problems and opportunities?
IV	Work	• Transactions • Level 3 and 4 processes • Best practice analysis	Final analysis: what solutions will be applied?
V	Implement	• Opportunity analysis • Project definition • Deployment organization	How, when and who will deploy the project?

Figure 7.11   SCOR project roadmap.

ONLY CEOS CAN BUILD AN ENTERPRISE CUSTOMER VALUE DRIVEN OPERATIONAL MODEL

I was Managing Director of 3M Brazil from 1995 to 2000. Together with my leadership team, we transformed the company to a true market segment and customer driven company. After three years, we had made great strides and completely changed the mindset and processes from selling products to meeting customer needs in the selected segments with complete solutions. The supply chain was also transformed as processes were rebuilt to align completely with our commitment to treating the best customers in each segment better. Although the supply chain director was one of the best leaders I have worked with, this would have been impossible if it was just a "supply chain initiative." Step function supply chain improvement requires that the CEO, the only person who really has end-to-end leadership, be the personal champion.

Paul Husby

---

and this should be included during project scope and time frame development. It is advisable for organizations who have not used SCOR as their roadmap for supply chain improvement projects to start with a narrow pilot project focused on a particular level 2 area such as make or deliver. Selecting these operational areas has the advantage of making improvements visible to the entire organization. These projects will identify shortcomings in processes like planning or scheduling and produce real measurable operational improvement.

Supply Chain Improvement Example: MMC

Strategy-Driven Supply Chain Design, Operation, and Improvement

In Chapter 3, four rules of a sustainable competitive business from Treacy and Wiersema's book, *Discipline of Market Leaders*,[2] were introduced, making an argument for establishing a direct linkage from business strategy to supply chain design, operational focus, and performance targets, and reinforcing continuous improvement as the right way to sustainably improve a company's competitiveness. That means applying their four rules:

Rule 1: Provide the best market offering by excelling in a specific dimension of value, which means performance excellence must be achieved in supply chain performance attributes most critical to success of a chosen value discipline and proposition.

Rule 2: Maintain threshold standards on other dimensions of value to ensure that they are not a cause of competitive disadvantage.

Rule 3: Dominate your market by improving year after year.

Rule 4: Build a well-tuned operating model dedicated to delivering unmatched value, which speaks directly to applying the LEOMS to design and operate supply chains. This infers supply chains must possess inherent attributes to deliver a business's value proposition routinely and consistently. Successful companies making up markets normally have specific supply chain priorities in support of their basic business strategy, model, and operational processes.

Companies will typically adopt one of three generic strategies:

1. Lowest cost, focusing on total cost of their supply chain from their suppliers through their own operations and delivery to their customers
2. Product innovation strategy focusing on differentiated product performance based on patents and trade secret process technology that requires more agility of suppliers and higher selling prices to deliver high margins and returns on their capital employed
3. A customer intimacy strategy focusing on being attuned to satisfying stated and unstated desires of their customers

Each of these generic strategies requires specific supply chain designs, relevant attributes, and appropriate metrics in support of the strategy's value discipline and proposition. The remainder of this chapter presents a strategy-driven Lean supply chain design, operations, and improvement example: the Minnesota Manufacturing Company (MMC).

MMC Business Situation Analysis

1. MMC was losing market share and profitability at an accelerating pace.
2. There was no clear strategy aligning customer value and all aspects of company operations.
3. Each function operated independently to achieve its goals.
4. The executive team respected each other and seemed to get along well, but they lived in a cordial anarchic environment where everyone treated each other respectfully but let them do their own thing; true teamwork did not exist.
5. The company had a long history of technical excellence and innovation but seemed to have gotten off track in development of new product and process technologies that added value to customers.
6. The company needed significant change to turn market share losses and declining profitability around.

MMC Value Discipline and Proposition-Driven Supply Chain Design, Operations, and Improvement

MMC's product leadership value discipline and product innovation value proposition (see Figure 7.12) requires advantage in delivery reliability, parity performance in supply chain responsiveness, advantage performance in supply chain adaptability/flexibility, and superior performance in cost of goods sold and asset management performance. The purpose of defining priority metrics is to get clarity and agreement on business and supply chain attributes and their related metrics to effectively drive supply chain design and operation. An understanding of a business's strategy is critical to effective supply chain attribute prioritization and metrics selection. It is the first step in enabling practitioners to be in position to begin designing/redesigning their supply chain to better deliver their company's value proposition and create sustainable continuous improvement. Examination of reality starts by establishing a clear picture of current financial and operational performance along with the value disciple and proposition, differentiating competencies, strategic processes, and priority metrics. Since the scope of SCOR assessment is the entire supply chain and all of its processes, designing

Value discipline – Product leadership
Value proposition – Product innovation

Operation model

> Business structure
> > Industry/market facing sales, marketing and technical service organizations
> > Product focused business units and teams
> > Strong product marketing, sales, applications, technical service and product and process development organizations
>
> Management systems
> > Market and product P&L structure
>
> Culture
> > New technology and product heroes
> > Product P&L gross margin drives decisions
> > Technology knowledge sharing and mentoring
> > Growth through new products and expanded applications of existing products and technologies

Differentiating competencies

> > Relationships with key customer technical management and knowledge of their products and services
> > Dedicated technical service organizations supporting sales representatives and customers
> > Knowledge of customer design, development and production processes, and key personnel
> > Participation in customers' product development process
> > Relationships with customer operations management teams
> > Materials science research
> > Future mapping of market and product trends
> > Technical community knowledge sharing
> > Proprietary process equipment design

Strategic processes

> > Product design, development, and commercialization
> > Process design, development, and commercialization
> > Intellectual property management process
> > Materials technology research and development
> > Technology and application intellectual property
> > Global technology scanning, evaluation, and acquisition
> > Manufacturing process design, proprietary equipment design, and operations management

Priority metrics

> > Product category penetration
> > Percent of sales from new products
> > Sales growth
> > Order delivered on time and in full
> > Product quality (DPPM)
> > Upside flexibility
> > Cost of goods sold percent
> > Return on capital employed

Figure 7.12   Product leadership value discipline.

and improving Lean supply chains starts by defining individual supply chains within their business. This is done by creating a matrix of product families and customer groups (see Figure 7.13). Segments or "customer groups" represent common expectations from their suppliers. Defining supply chains is a first step in effectively establishing "voice of customer" requirements and associated enabling supply chain business processes to be aligned with these requirements.

Which of the three supply chain types does MMC have? MMC's supply chain is a make-to-stock supply chain shown in Figure 7.14, so level 2 processes are S1, M1, and D1.

MMC sells high-volume products designed for each of their medical device customers, medical device distributors who supply replacement parts, and dealers

MMC supply chains

Customer groups ➡ Product families ⬇	Medical applications	Sports applications	High-tech applications
Mechanical assemblies	90% of revenue	2%	3%
Metal components	4% of revenue	.4% of revenue	.6% of revenue

Figure 7.13   MMC supply chains.

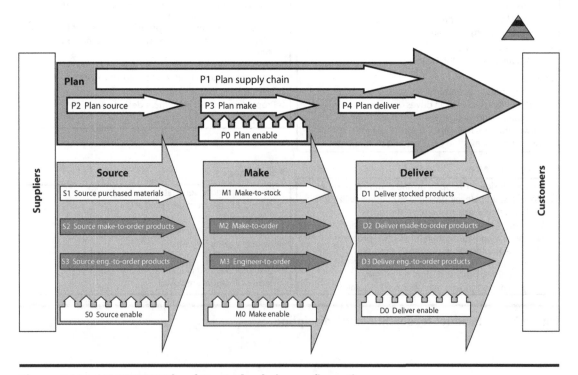

Figure 7.14   MMC SCOR level 2 supply chain configuration.

LESSON LEARNED

3M floppy disk manufacturing was an intensely competitive business requiring a sustained unit cost improvement rate of 15%–20% per year for ten years. While improvement efforts were primarily focused on cost, growth and satisfied customers were also necessary conditions for survival. Entering two new markets, lessons about different market segments service and quality expectations were learned the hard way. Sourcing 3M Japan was the first new market to supply as their manufacturing didn't have sufficient volume to be successful. When volume shipments began to 3M Japan, a high percentage of product was rejected for quality reasons. These were the same floppy disks supplied to U.S. markets, so the organization was stumped regarding the source of the problem. The difference turned out to be expectations. Eventually success in Japan was attained, but only after establishing a formal quality improvement team and addressing issues in monthly teleconferences with our colleagues in Japan. The second market was the computer OEM (original equipment manufacturers) market. These customers valued quality and delivery performance above cost but, our plant processes weren't geared to respond efficiently. They were incredibly demanding about exactly the right color on packaging materials at a level the human eye couldn't detect, to meet their quality expectations, including delivering perfectly. The point is both 3M Japan and OEM's represented different customer groups with different priorities. Success was achieved by managing these "supply chains" differently to meet their expectations.

Paul Husby

---

who sell products to startup and small device manufacturers. They also sell product to sports fitness monitoring equipment and high-tech control system equipment companies, both growing markets with great growth potential.

The first step, supply chain attribute prioritization, is accomplished by determining each market's priorities using the Lean supply chain metric prioritization form shown in Figure 7.15.

Once supply chains are defined for each customer segment, a balanced set of supply chain metrics is needed to create value for customers and financial return for MMC. Many companies supply products into multiple markets that may well place value on different supply chain attributes, so it is important to reflect each of their markets' priorities in the design. These metrics must reflect "voice of customer" priorities along with operational and financial performance improvement, resulting in a balanced internal and external picture of a business's operational health. With key attribute priorities defined for each major market served, the next step is to determine MMC performance requirements to achieve its strategic goals and define current performance gaps.

MMC supply chain metric prioritization

Performance attribute	Prioritized performance needed to win		
	Medical device	**Sports and fitness**	**High technology**
Delivery reliability	Advantage	Advantage	Advantage
Responsiveness	Parity	Parity	Parity
Flexibility/agility	Advantage	Parity	Advantage
Supply chain cost	Superior	Superior	Superior
Return on assets	Superior	Superior	Superior

Figure 7.15   Lean supply chain metric prioritization.

Value Proposition and Strategy Drive Operational Attributes and Metric Dashboard

Creating Supply Chain Dashboards

Few activities will be as important to increasing long-term strategic competitiveness through redesigning and continuously improving supply chains as selecting level 1 metrics that will provide a long-term supply chain improvement focus. Too often, short-term problems and the appeal of reducing cost impede efforts to continuously improve long-term strategic competitive advantage. Doing this right dictates leaving aside organizational and individual paradigms, current metrics, and opinions. The goal is to create a dashboard that

1. Enables organizational clarity and understanding of metric linkage to continuously increasing long-term company competitiveness
2. Defines performance targets aligned with strategic and operational business objectives
3. Provides a balanced view of customer-facing and internally facing metrics
4. Is organized to measure entire supply chain performance
5. Is linked to action; metrics that reveal actual and target performance along with deviations from target performance

The SCOR model supports creation of a meaningful dashboard by providing a list of cross-industry performance attributes to guide metric selection. These are

1. Delivery reliability—Reliability in delivering product on time and in perfect condition
2. Responsiveness—How quickly orders are fulfilled and shipped
3. Flexibility—Ability to respond to unplanned demand

4. Cost—Product production cost or total supply chain
5. Asset management—UtilAgility/flexibilityization

Using SCOR performance attributes as a guide (see Figure 7.16), at least one metric per category is established. An analogy would be an automobile dashboard containing many gages and lights but missing an oil pressure indicator. Everything seems to be running fine until low oil pressure results in engine failure. Likewise, metrics or key performance indicators need balance to ensure that all important areas of performance are in focus. Once sources of customer value are understood, metrics aligned to those sources and their required performance standards can be defined and validated by customers. MMC chose SCOR model metrics as shown in Figure 7.17; however, possible metrics for each category are not limited to this list and may include level 3 metrics. For example, in the category of delivery reliability, typical choices might include:

■ On-time shipping.
■ Fill rate.
■ Perfect order fulfillment (SCOR)—very difficult to measure routinely but is excellent to use for an in-depth study every couple of years to be sure that something is not being missed. It is best done by selecting a set of customers in a major market segment to be used as the data source.

A useful guideline when identifying metrics is to set aside constraints and seek to define the best metric possible, one that accurately measures customer expectations and the most important internal performance requirements. With the metrics selected, data collection methods must be considered. Measurement constraints may appear making it necessary to

MMC medical market strategic metric current state and targets
Strategic targets: MMC performance improvement required to win?

Attribute	Strategic metric (level 1)	Current state	Performance targets		
			Parity	Advantage	Superior
Reliability	Perfect order fulfillment	Disadvantage		▓	
Responsiveness	Order fulfillment cycle time	Disadvantage	▓		
Agility/Flexibility	Supply chain flexibility	Disadvantage		▓	
	Supply chain adaptability	Disadvantage			
Cost	Cost of goods sold	Superior			▓
Assets	Cash to cash cycle time	Disadvantage			▓

Figure 7.16 MMC strategic metric targets.

Strategic targets: What metrics will be used to measure success?

Attribute	Strategic metric (level 1)	Selected operational metrics	Parity	Advantage	Superior
Reliability	Perfect order fulfillment	Percent of orders shipped on time		▓	
Responsiveness	Order fulfillment cycle time	Order fulfillment cycle time	▓		
Agility/ Flexibility	Supply chain flexibility	Percent equipment utilization (24 × 7)		▓	
	Supply chain agility	Total production lead time days		▓	
Cost	Cost of goods sold	Cost of goods sold			▓
Assets	Cash-to-cash cycle time	Days of inventory			▓

Actual: Your organization's performance

✓ Parity: The performance of the middle company (median)

✓ Superior: The performance of the top 10 percentile performer(s)

✓ Advantage: The middle of parity and superior performance

Figure 7.17   MMC operational metric targets.

compromise. For example, a metric such as "on-time delivery" may be difficult as actual delivery times may not be available, so it is necessary to substitute "on-time shipping," which can be measured internally. As noted in Chapter 6, MMC found that they had only been measuring from order release to shipment, so they used the data they had and began collecting data on order entry to release lead time. MMC chose a combination of level 2 and 3 metrics based on their need to improve strategic advantage by improving customer service metrics and operational advantage by creating additional cash to invest in technology.

Reviewing Strategy

SCOR Benchmarking Results

Benchmark data are divided into quartiles (Figure 7.18):

1. The bottom two quartiles of performance indicate competitive disadvantage.
2. Median performance 3rd quartile, 50th percentile to 74th percentile is neither advantage nor disadvantage.
3. Advantage performance is in the 4th quartile of the 75th to 89th percentile.
4. Best in Class performance is in the 90th percentile and above indicating superior performance versus competition.

The MMC results were color-coded—black is disadvantaged, dark grey is at median, light gray is advantage, and white is best-in-class; unfortunately, black

MMC Benchmark Performance Generation 1	On time shipment performance percent	Line fill rate percent	Order release to shipment lead time	Total production lead time	Cost of goods sold percent	SGA cost percent	Warranty/returns cost percent	Cash to cash cycle in days	Days of inventory	Asset turns	Gross margin	Operating income percent	Net income percent	Value added per employee ($ 000)	Return on capital employed	3 yr CAGR percent	Revenue ($ 000,000)	Number of employees
MMC	94%	94%	7	47	66.2%	21%	6%	62	47	6	33.8%	12.8%	8.7%	$31	11.3%	6%	$280	775
Company A	83%	95%	4	59	75.0%	18%	9%	80	58	4.0	25.0%	7.0%	4.8%	$13	7.5%	0%	$140	500
Company B	92%	93%	3	64	72.0%	17%	6%	65	63	6.0	28.0%	11.0%	7.5%	$22	13.0%	5%	$320	1100
Company C	96%	98%	3	58	78.0%	13%	4%	55	57	7.0	22.0%	9.0%	6.1%	$21	11.6%	8%	$285	825
Company D	87%	81%	4	74	78.0%	20%	10%	90	71	3.6	22.0%	2.0%	1.4%	$5	2.4%	–3%	$700	2100
Company E	91%	94%	2	49	71.0%	15%	5%	50	46	3.1	29.0%	14.0%	9.5%	$26	12.0%	7%	$210	760
Median	91.5%	94.0%	3.5	58.5	73.5%	17.5%	6.0%	63.5	57.5	5.0	26.5%	10.0%	6.8%	$21	11.5%	5.5%	$283	800
Advantage	93.5%	94.8%	3.0	51.3	71.3%	15.5%	5.3%	57	49.5	6.0	28.8%	12.4%	8.4%	$25	11.9%	6.8%	$311	1031
Best in class	95.0%	95.0%	2.5	48.0	68.6%	14.0%	4.5%	52.5	46.5	6.5	31.4%	13.4%	9.1%	$29	12.5%	7.5%	$510	1600
Gap to median	-2.5%	0.0%	5.0	-11.5	-7.3%	3.5%	0.0%	-2	-10.5	-1.0	-7.3%	-2.8%	-1.9%	-$10	0.2%	-0.5%	$3	25
Gap to advantage	-0.5%	0.8%	4.0	-4.3	-5.0%	5.5%	0.7%	5	-2.5	0.0	-5.0%	-0.4%	-0.3%	-$6	0.6%	0.8%	$31	256
Gap to best in class	1.0%	1.0%	4.5	-1.0	-2.4%	7.0%	1.5%	10	0.5	0.5	-2.4%	0.6%	0.4%	-$3	1.2%	1.5%	$230	825

Figure 7.18   MMC benchmark summary.

and dark grey dominated the MMC cells. Management teams are often stunned and question data validity, and may even complain that this process is intended to present their results in the most negative light or agree among themselves that last year was just a bad year. Eventually, groups faced with results that counter their own perception of their operational excellence will accept brutal reality; in MMC's case, performance is median or less across a majority of measured parameters.

Analyzing Competitors

MMC's competitors' strategies were evident in analyzing their data. Competitor C chose operational excellence as their core strategy, building an operational model to provide a superior level of delivery performance very efficiently. Benchmark results (Figure 7.18) validated their strategy as they had superior performance in delivery reliability. Competitor E chose customer intimacy as their core strategy built on operating model with superior responsiveness and flexibility and a very broad product line to satisfy the needs of their customers. While they manufactured many of the high-volume products, many of the B and C volume items were purchased. MMC understood why they had been losing market share. Their top two competitors had chosen unique strategies, supported by well-designed operating models ensuring effective operational implementation. MMC had become complacent and had not recognized performance gaps that evolved over time, resulting in their current weakened competitive position after years as the market leader based on their *product leadership strategy*. They were still selling innovative products that provided differentiated value but not sufficiently to overcome their high selling price combined with their poor service. Fortunately, the implementation of the LEOMS over the previous year started them on a path to restore their historical market leadership position as a result of improvement of their key supply chain metrics performance:

- On-time shipping percent improved from 90% to 94%—from disadvantage to median.
- Order release to shipment lead time—even after improvement, their order filling and shipping lead time improved from 4 to 2 days and but they remained disadvantaged as they had not recognized the 5 days of order entry to order release lead, time assuming it was 2 days.
- Total production lead time improved from 82 to 47 days—from disadvantage to advantage performance.
- Cost of goods sold improved from 68% to 66.2%—increasing superior performance.
- Asset turns improved from 3.5 to 6 turns—from disadvantage to advantage performance.

Determining Parity, Advantage, and Superior Performance

Benchmark results are calculated and presented in Figure 7.19. Parity is determined by locating the median data point. Superior is performance in the top 10 percentile. Advantage is the midpoint between the parity and superior values. Actual performance is entered in the metric table, and gaps are determined by comparing them to desired performance based on each metric's priority. Gaps to achieve desired performance should be integrated into strategic and operational planning along with strategies and operational plans to close them.

Business strategy aligned supply chain performance		Strategic value discipline – proposition		
		Competitor C	MMC	Competitor E
		Operational excellence	Product leadership	Customer intimacy
		Lowest delivered cost	Technological innovation	Total solutions provider
Customer facing	**Delivery reliability**	Superior	Median	Median
	Responsiveness	Advantage	Disadvantage	Superior
	Adaptability/flexibility	Median	Superior	Superior
Internal facing	**Cost**	Disadvantage	Superior	Superior
	Asset management	Advantage	Advantage	Disadvantage

Figure 7.19   MMC supply chain competitive strategy assessment.

Financial and Operational Benchmarking Sources

Once supply chain metrics are identified, benchmarking can proceed. Financial benchmarking of publicly traded competitors and market peers is most commonly done using their 10K reports available from various benchmark services such as Hoovers and Forbes. Data of privately held competitors are not reported publicly and therefore are not included in available data sources. Data are also available from free sources but only at the company level, so comparison to multi-division competitors is not possible. The company's chief financial officer can provide these benchmark comparisons from available sources. Financial benchmarking builds a business case by gaining insight into supply chain improvement opportunities. Operational benchmarking data gathering may require multiple sources to get a comprehensive picture. American Productivity and Quality Council (APQC), a free service for Supply Chain Council members, is an excellent source of benchmarking data. APQC provides industry-level data but has limited capability to provide data below the industry level. Other data sources include trade associations, published articles and reports, benchmarking services, and customers. Customers normally evaluate suppliers using common criteria. They compare market peers to standards established by the best supplier performances, so not having direct competitors in a benchmarking process does not diminish its value, as the goal is satisfying customers.

MMC Market Situation Analysis

MMC intentionally waited until they had some improvement results before they visited customers as they were aware they had little credibility because of their noncompetitive operational performance. MMC's customers continued to value their innovation but, in recent years, had increasing demands for operational excellence in delivery and responsiveness that MMC had not met. Their conversations with key customers all followed the same observations regarding their improvements and remaining shortcomings. Customers recognized MMC's on time and line fill rate had improved, and they had become more agile, responding quicker when customers had demand spikes. But MMC's order fulfillment lead time still was noncompetitive. MMC did not need to change their fundamental strategy but needed to sustain superior performance in product innovation and at least threshold performance in other attributes of operational excellence, which they have done except in the case of the order fulfillment lead time. This defined MMC's challenge in eliminating the remaining competitive disadvantage to enable restoration of their credibility and their product leadership. MMC had started their LEOMS journey with a focus on factory operations and, after the first year, fully engaged their supply chain and customer service professionals to accelerate source, plan, make, and deliver improvement.

Creating Value for Customers

Success of any strategy requires superior satisfaction with target customers. Understanding customer needs, expectations, and priorities requires a depth of information, which can only be obtained through a continuous relationship, a collaborative partnership that is beneficial to both parties. Identifying a segment of customers and building a deep relationship so that they want to provide service and product quality information is very important to ensuring that business strategy actions are aligned with customers' real needs and priorities. These customers are also likely to advise the supplying company when things are starting to go wrong.

Gathering Customer Information

A plan was needed for efficiently gathering customer information. The key questions to answer first were as follows:

1. What information is needed?
2. Which customers shall we approach?
3. What process will we use?

After much discussion, they defined four key pieces of customer information that must be understood:

1. Their business sourcing metric priorities
2. Their view of MMC's performance including data if available
3. An assessment of how MMC ranked against competition
4. Specific issues the customer had with MMC

Data desired from each selected customer included

1. Customers' measurement of MMC's on-time delivery
2. Customers' measurement of MMC's delivered quality
3. Customers' measurement of MMC's perfect order fulfillment
4. Customers' experience with MMC's order fulfillment lead time

MMC knew that it would be impossible to approach all customers. MMC created a customer list from largest to smallest based on sales and selected three large customers that were critically important to their business (Figure 7.20). Next, they reviewed the remaining list of customers, searching for other strategically important customers. Results from this search added one company for a total of four candidates to visit and gather needs and priorities. Using five

MMC actual performance	1st priority	2nd priority	On-time delivery	Delivered quality	Other delivery issues	Perfect order fulfillment	Order fulfillment cycle time
LeBlax	Delivery reliability	Responsiveness	92%	97.4%	95%	80.8%	16 days
MexAms	Responsiveness	Delivery reliability	89%	90.3%	97%	76.1%	14 days
Jonso Medical	Delivery reliability	Responsiveness	88%	94%	94%	79.8%	12 days
Zyxain Med.	Delivery reliability	Responsiveness	90%	90.2%	96%	80.9%	13 days
Overall	Delivery reliability	Responsiveness	89.7%	93.0%	85.5%	79.4%	13.8 days

Best supplier performance	On-time	Delivered quality	Other delivery issues	Perfect order fulfillment	Order fulfillment cycle time
LeBlax	98%	88%	99%	85%	9 days
MexAms	92%	95%	97%	75%	8 days
Jonso Medical	95%	93%	95%	80%	7 days
Zyxain Med.	91%	97%	95%	85%	9 days
Benchmarks	94%	93.3%	96.5%	81.3%	8.25

Figure 7.20 MMC customer priorities.

SCOR performance attributes, several questions were developed to pose to key customers.

1. Of these five performance attributes, which is your top priority and which is second?
2. How does MMC rank vs. similar suppliers in each of these five customer measures?
   a. On-time delivery
   b. Delivered quality
   c. Product performance
   d. Perfect order fulfillment
   e. Order fulfillment lead time
3. For each performance category, what performance level is required to be your best supplier?

MMC executives visited customers and learned that they continued to value MMC's innovation and new products, but their service on established products was not acceptable and hurting their customer's business. In addition, competitors had shorter order fulfillment lead time, allowing customers to carry lower inventory.

Year 2 Business Plan

MMC Business Current State

MMC needed to accelerate regaining market leadership and sales growth to increase profits that will accrue from regaining their historical market position. They realized how market leadership could be reestablished and sustained using this assessment methodology to guide the journey. They also understood that competitive benchmark performance is not static as their competitors are always improving, so achievement of an advantage position may not be enough if competitors are improving at a similar or faster pace.

MMC Year 2 Operational Improvement Plan

Year 2 targets needed to progress toward closing identified supply chain performance gaps to reestablish market leadership. MMC top executives spent 2 days together reviewing results and determining their path forward by adapting the annual plan to include the targets established from benchmarking. Transformational change was required and would be an enormous task and take 2 to 3 years. After two long days of discussion and debate, they agreed on the following assumptions and goals:

1. MMC commits to growing faster than market average in the second year of LEOMS implementation by
    a. Achieving advantage on time shipment performance, capacity utilization, and cash-to-cash cycle time performance
    b. Improving order fulfillment lead-time performance from major opportunity to disadvantage
    c. Sustaining their superior supply chain cost position
    d. Improving net asset turns to median performance
2. The primary approach to achieving service goals will be achieved through aggressive LEOMS application to their entire supply chain.

MMC's leadership team committed to pool all existing improvement resources and generate additional project resources from early improvement as resources cannot be added now due to current financial performance.

Business Plan Operational and Financial Targets to Improve

Achieving the vision would mean regaining market leadership, satisfy their customers, and dramatically improve operational performance. The MMC's leadership team quantified the annual operational plan targets (see Figure 7.21).

Year two supply chain annual targets to improve																		
Each target you have previously entered will appear below automatically. For each target, enter your starting point and plot the progression needed to meet the target on time. Fill in the actual progression over time. Current cell formatting will show																		
Project status	Last status review date:			Green	On-time	Yellow	At risk	Red	Off-plan	Initiate		Continue						
Priority	Deliverables and measures	Starting point	Month	1	2	3	4	5	6	7	8	9	10	11	12			
1	Supply chain reliability – improve percent of order shipped on time to 98%	94%	Plan															
			Actual															
2	Supply chain responsiveness – improve order fulfillment L.T. average to 4 days	7 days	Plan															
			Actual															
3	Supply chain flexibility – reduce 24/7 utilization by 10 percent to 88.5%	96%	Plan															
			Actual															
4	Supply chain agility – reduce total production lead time to 40 days	47.1 days	Plan															
			Actual															
5	Reduce cost of goods sold to 65%	66.2%	Plan															
			Actual															
6	$10 M increase in cash by reducing days of inventory to 40	47 days	Plan															
			Actual															

Figure 7.21　MMC supply chain annual targets to improve.

1. Improve on time shipping performance to 98%
2. Reduce order fulfillment lead time to 4 days
3. Reduce capacity utilization to 88.5%
4. Improve supply chain agility by reducing total production lead time to 40 days
5. Reduce cost of goods sold to 65%
6. Increase cash by $10 MM by reducing days of inventory to 40 days

The plan's projected financial benefit increases operating profit by $13 MM and a cash improvement of $19 MM, and a 1.2 turn asset turns improvement.

Expected Competitive Improvement

The bottom line was MMC's product leadership strategy had become ineffective due to poor service, resulting in business gained with new products being quickly lost to competitors. Their best competitors follow fast with new similar products, although seldom with the quality and performance of MMC products, but with their superior service, they were frequently replacing MMC new products. Benchmark results validated by customers made it clear that MMC's operational system must be redesigned to consistently meet delivery reliability, quality, and responsiveness expectations of their customers. In Figure 7.22, benchmark results are presented in five columns, which divide data populations into these categories:

1. Column 1: bottom quartile of performance and a major opportunity.
2. Column 2: second quartile of performance is a competitive disadvantage.

MMC supply chain year two targets

Key perspectives	SCOR level 1 metrics	Supply chain performance versus custom population					Plan
		Major opportunity	Disadvantage	Average or median	Advantage	Superior	
Delivery performance and product condition	On-time shipment percent		☞	●	☆	△	+5% Sales Growth = $13 M yr. 1
	Perfect order fulfillment	☞		●	☆	△	
Flexibility and responsiveness	Capacity utilization		☞	●	☆	△	≤88.5% Util. Offset LEOMS start up cost
	Order fulfillment lead time	☞	☆	●	◆	△	4 Days
Cost	Supply chain management cost			●	◆	☞☆	−1.2% Points $4.6 M/yr.
Working capital	Cash to cash cycle time		☞	●	☆	△	7 Day reduction
	Net asset turns		☞	☆	◆	△	.2 Turns $9.3M Inv. reduction.
Total benefit included in S.C. management cost		First full year operating profit benefit					$13.0 M
		One time inventory reduction					$10 M

☞ MMC Position   ● Median   ◆ Advantage   △ Superior   ☆ Targets

Figure 7.22   MMC SCOR benchmark supply chain targets.

3. Column 3: average or median performance.
4. Column 4: 75th percentile of performance is a competitive advantage.
5. Column 5: 90th percentile of performance is a superior position versus competition.

MMC benchmark results defined performance gaps from current levels to their chosen target for each metric. The column on the far right quantifies potential benefits of achieving these performance levels, building a business case for change. MMC executives realized that this assessment provided a clear picture of current performance and gaps that are aligned with their business strategy and customer priorities. They realized how this information would mobilize their organization once they communicated these results broadly across MMC.

Improving Supply Chain Performance to Meet the Targets

Supply Chain Current State

The supply chain vice president's team of directors responsible for manufacturing, sourcing, planning, and logistics were established as the leadership team responsible for improving process planning and execution along with adapting their organization's culture to ensure improvement methodology sustainability. Having completed their value stream map and revised their targets (see Figure 7.23), reflecting sourcing, logistics, planning, customer service and order

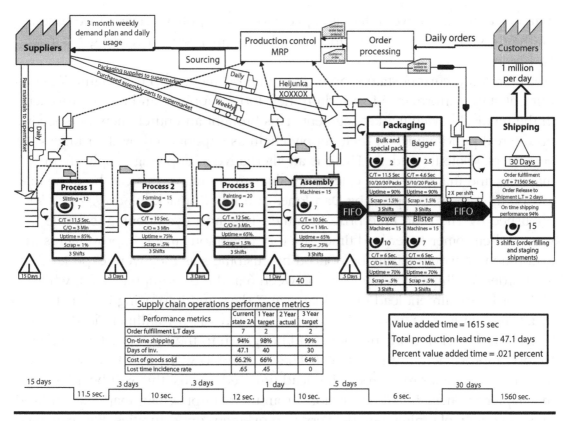

Figure 7.23   MMC generation 2A current state value stream improvement map.

processing operations on the Value Stream Map. They had always assumed that the total lead time from order entry to order release for picking was 2 days instead of the 5-day actual lead time, but this would have to wait as it would be addressed in their next phase of LEOMS implementation when transactional process continuous improvement is introduced in Chapter 8.

Diagnosis—Find Opportunities to Improve

Having documented feedback from key customers, MMC was now ready to make an internal operations assessment to define needed improvements and set improvement targets. Having started implementation of the LEOMS a year earlier, MMC factory operations had started improving their shop floor operations and now needed engagement of sourcing, planning, logistics, and customer service to improve the entire supply chain.

Documenting the Supply Chain Current State

All successful businesses segment their customers into groups, each group having its own unique needs and expectations. For example, mass merchandisers like Wal-Mart and Target have very different supplier expectations than original equipment manufacturers, distributors, government customers, or others. Within markets, customers have different expectations of supplier performance based on their own business strategy and related performance metrics. To get started, MMC's logistics organization worked with sourcing and customer service to create an MMC current state supply chain improvement map (see Figure 7.24). Current state performance focused on their top 3 suppliers and top 4 customers in the medical device market segment. This brought an entirely new perspective on the real performance of their supply chain as experienced by their biggest customers. (Note to readers: There is an alternative to applying supply chain mapping and that is to draw extended value stream maps; this can be done following the process supplied by Dan Jones and Jim Womack in *Seeing the Whole*[3] published by The Lean Enterprise Institute.) The most eye opening example was total customer order lead times. Factory operations made improvements in lead time that would hardly be noticed by customers as they were experiencing total order fulfillment lead times of between 6 and 8 days suppliers. So, while MMC order fulfillment lead time alone was 7 days and improving it by 1 day meant that customers would still experience 6-day order fulfillment lead times. The customer service organizations had bought in to the LEOMS after seeing the factory improvement in the last year. They wanted to work with factory operations to implement order processing lead-time reductions; this will be addressed in Chapter 8. These initial conclusions from their supply chain map data showed great sourcing, planning, and delivery improvement opportunities. For example, in Figure 7.25, the level 2 metric of source cycle time is affected by the performance of three-level process elements that are measured by related metrics:

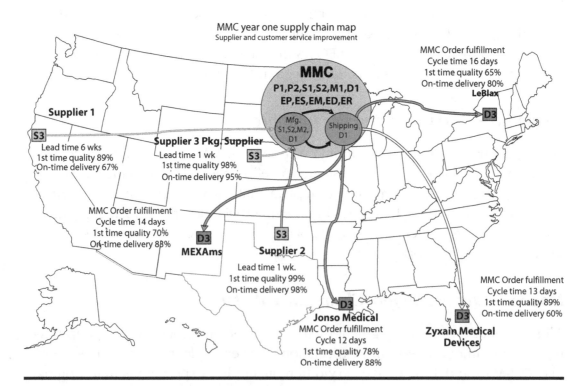

Figure 7.24 MMC current state supply chain map.

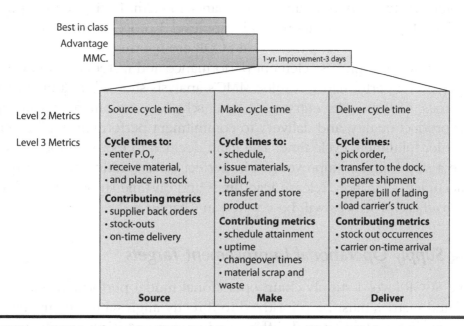

Figure 7.25   MMC total supply chain lead time decomposition analysis.

cycle times to enter purchase orders, receive material, and place the material in stock. Digging deeper into back orders would include a measurement of items and suppliers that are most common cause of stock item zero balance occurrences. Effective diagnostic metrics are of great value in being able to quickly identify the source of average source cycle time and its deviation, giving clarity of focus to the most significant deviations and their corrective actions. Based on customer priorities, MMC's initial focus would be on "deliver cycle time," as this was totally within their control and customers would feel immediate impact from reduced delivery times. They created the perfect order fulfillment process model to understand which processes contain root causes of supply chain performance gaps (see Figure 7.26). This is the power of defining and prioritizing supply chain attributes and SCOR level 2 metrics, reflecting what is important

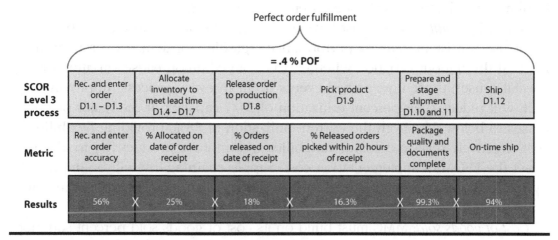

Figure 7.26   MMC perfect order fulfillment.

to customers: internal performance and financial health. Reviewing their level 2 metrics led MMC to identification of relevant level 3 processes and metrics, shown in Figure 7.26. Level 3 metrics in SCOR are referred to as diagnostic metrics as they measure process element performance and factors that in aggregate make up level 2 performance results. MMC's analysis showed multiple process failure areas, such as order entry cycle time, schedule attainment, packaging quality, product quality, and delivery to commitment performance, all contributing to order fulfillment cycle time. The team decided to focus on four processes to make a step function improvement in perfect order fulfillment: order receipt, entry accuracy, allocation, release lead times and on-time shipment. Applying LEOMS to these processes will be covered in Chapter 8.

Setting Supply Operational Improvement Targets

The five SCOR level 1 supply chain operational model performance attributes and related metrics must be prioritized to directly align with a company's value proposition, strategy, and differentiating competencies. This focuses Lean supply chain operational model design parameters and continuous improvement on the greatest potential gains in customer satisfaction and shareholder value. For each of these five performance attributes, a metric is chosen to support the value proposition and be the focus of continuous improvement to achieve MMC's 3-year strategic targets.

Delivery reliability: To sustain customer confidence and loyalty, MMC must do more than supplying differentiated products by also demonstrating their commitment to the highest standards in quality, reliability, and service by improving on-time shipments by 98% in year 1 and 99% in 3 years to achieve an advantage performance position in perfect order fulfillment.

Responsiveness: MMC's order fulfillment cycle time will be improved to 6 days in the next improvement cycle as order release to shipping would improve to 1 day, and order entry to order release improvement would be undertaken in the third year of the improvement plan to ultimately achieve a 2-day total order fulfillment lead time; this will be covered in Chapter 8.

Agility/flexibility: MMC must have advantaged flexibility and adaptability to ensure consistent delivery performance regardless of demand variation or internal throughput variation whether it is from common cause variation or special causes; those unexpected events occur in every process. Improving OEE will reduce their percent utilization of 24/7 capacity of the major processes to below 75% from the current utilization of 88%, providing flexibility to cost-effectively service demand variability and improve productivity to at least offset the LEOMS's implementation cost. Supply chain agility will be improved by reducing total production lead time to 40 days in 1 year and to 30 days over 3 years.

Cost of goods sold: MMC must build on its cost of goods sold percent superior position by reducing unit cost while improving quality and service, providing customers with increased value that more than justifies their selling prices. This is critical to achieve MMC's strategy as they must continue to increase investment in basic materials and process research along with product and process development to sustain their product innovation leadership position.

Cash-to-cash cycle time: MMC must achieve a higher return on capital employed to achieve a superior position by sustaining the price premium and efficient management of fixed assets, inventories, accounts payables, and accounts receivables. This will be achieved by reducing 24/7 utilization to 75% or less without additional capital investment in equipment and improve cash-to-cash cycle time by reducing days of inventory to 40 days in the next year and 30 days in 3 years. Enterprises in MMC's position frequently are shocked and in denial when their current performance and "required performance to win" are communicated to their organizations. This response is common in organizations with innovative products protected by patents and proprietary process secrets as they wrongly interpret their excellent financial results as proof of their excellence in supply chain competence and performance. Like all closely held myths, organizations will hold on to these false beliefs even when financial performance is declining to a point of threatening their survival, creating financial conditions that make it difficult to make significant technology and technical staff investments required to sustain product innovation a differentiation that has provided their company with excellent financial results.

Establishing Benchmark Targets

Setting strategic supply chain targets should be an integral part of strategic planning using knowledge of market trends, competitors, and benchmarking analysis. There are numerous sources of benchmark information listed in Appendix IV. Cost varies greatly from source to source; in general, costlier sources make it easier to obtain good comparable market data. Another good source of benchmark information is trade associations who frequently complete market studies and sometimes benchmark participant companies. A good place to start is simply gathering information from a wide range of customers as a very viable method for documenting benchmark performance. Benchmarking is not competitive analysis, as comparative populations should include a broad group of peer suppliers who serve the same markets. While a competitor may well be part of a benchmark population, measuring performance versus a wide range of top-performing peer companies is the best way to go in order to understand the levels of performance needed to be a top supplier. Getting started with benchmarking for target-setting purposes is most practically done by using existing company information because the first priority is to get started in making improvements. This process can be improved upon each year by refining formal benchmarking practices to gain greater external perspectives and data. Benchmarking should include all metrics selected for measuring all five supply chain attributes as a minimum.

Supply Chain Improvement Plan

While a 2-day order release to shipment lead time was achieved, their customer's overall order fulfillment lead time was only reduced between 9 and 12 days (see Figure 7.27), and it needed to be reduced more so they would become a leader in service with the shortest lead times and highest reliability. The delivery logistics organization had focused on improving their own warehouse operations but had not given much attention to order entry and order processing processes with its 5-day process lead time. This oversight would have to be a priority focus in the next year to reduce the excessive overall order fulfillment process lead times by applying business process mapping and the LEOMS's eight wastes to identify improvement opportunities (see Chapter 8). Supply chain planning and scheduling will focus on lot size reduction and in-process inventory reduction in collaboration with factory operations to continue to reduce overall supply chain lead time, which will improve both flexibility and agility capability of MMC's supply chain. To achieve their targets, operations is going to focus on improving process throughput, reliability, responsiveness, and cost by performing the following items:

■ MMC for the first time calculated OEE for every process, and the range was 29% to 69%; they had worked hard through implementation of 5S, work cell redesign, and changeover times, but OEE percentages remained at unacceptable levels, which means they had to put greater effort into preventative and predictive maintenance to improve uptime.

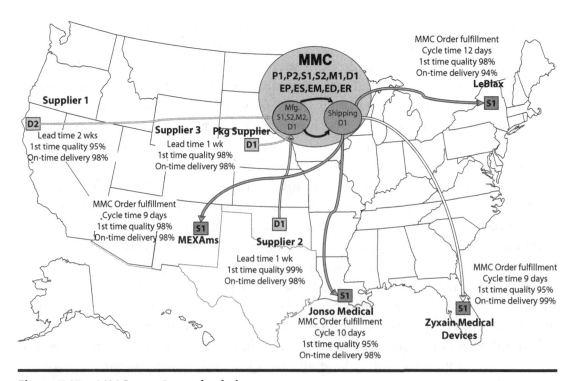

Figure 7.27   MMC year 2 supply chain map.

- Continue process engineering efforts to reduce scrap/waste by at least another 50% through process variability and scrap/waste reductions.
- Engage operators in a challenge to focus on work cell labor waste by identifying opportunities and working with their group leader to eliminate all waste by improving processes.
- Increase velocity by reducing in-process inventory to provide scheduling with more flexibility and increased capability to meet 99.7% of peak demand variability and reduce inventory to 40 days.
- Achieve 66% cost of goods sold a net 0.2% reduction including MMC's hiring of 15 new scientists to work on proprietary process technology and building a new process technology research and development facility; the operating cost of a process technology lab is included as part of factory cost.

MMC Year 2 Results

MMC supply chain leadership maintained their focus on achieving planned results and felt good about what their organizations had achieved in the first year. The LEOMS had been implemented, completing all seven waves, and now they were ready to repeat the cycle as there were still a lot of improvement opportunities and a few competitiveness issues that needed to be resolved. The sourcing and delivery logistics organizations had made solid progress improving their suppliers' delivery and quality performance. They also made progress on improving their own on-time shipments to 98%; see the major customer on-time delivery performance in Figure 7.28. Factory operations continued their aggressive assault on process variability to drive up OEE and up-times and achieve their 66% cost of goods sold target.

MMC Year 2 Operational Review

Reviewing their improved competitiveness in each of the six key supply chain attributes in Figure 7.29 showed that they had met or exceeded targets in four of the seven attribute metrics: capacity utilization, supply chain management cost, cash-to-cash cycle time, and net asset turns, while they failed to meet their on-time shipment percent, perfect order fulfillment percent, and order fulfillment lead time. The review team moved on to see how well they had done in improving the supply chain operation's contribution to their strategic market position.

Benchmark Review

The supply chain organization completed an update of the benchmark process and presented it to the LEOMS implementation leadership team and company executives (see Figure 7.30). Significant progress had been made, even though some of the targets were not achieved. Only two items remained in the category of disadvantage: order fulfillment lead time and SG&A cost percent, which is not

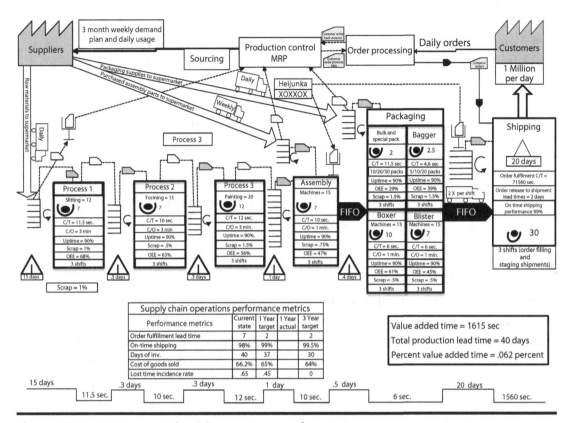

Figure 7.28   MMC generation 3 current state value stream map.

Key perspectives	Scor level 1 metrics	Major opportunity	Disadvantage	Average or median	Advantage	Superior	1 yr. Actual benefit
Delivery performance and product condition	On-time shipment percent			●	☞ ◆	☆	+9.4% Sales growth
	Perfect order fulfillment		☞	●	◆	☆	$30 MM Revenue increase
Flexibility and responsiveness	Capacity utilization			●	☞ ☆	△	88% Capacity utilization
	Order fulfillment lead time		☞	●	☆	△	1 days (14%) Reduction
Cost	Supply chain management cost			●	◆	☞ ☆	−1.8% COGS $5.2 M/yr. GM Increase
Working capital	Cash to cash cycle time			●	☆	☞ △	14 day reduction
	Net asset turns			☆	◆	☞ △	1.2 turns $20M Inv. reduction.
Total benefit included in S.C. management cost*		First full year gross profit benefit (gross margin % reduction and sales increase percentages above previous three year average)					$6.3 M
		One time inventory reduction					$19.3 M

Supply chain performance versus custom population

☞ MMC Position   ● Median   ◆ Advantage   △ Superior   ☆ Targets

Figure 7.29   MMC year 2 operational review scorecard.

MMC Benchmark Performance year 2	On-time shipment performance percent	Line fill rate percent	Order fulfillment leadtime	Total production lead time	Cost of goods sold percent	SGA cost percent	Warranty /returns cost percent	Cash to cash cycle in days	Days of inventory	Asset turns	Gross margin	Operating income percent	Net income percent	Value added per employee ($ 000)	Return on capital employed	3 yr CAGR percent	Revenue ($ 000,000)	Number of employees
MMC	98%	98%	6	40	66.2%	21%	6%	48	40	7.8	33.8%	12.8%	8.7%	$39	12.8%	6.5%	$350	775
Company A	83%	95%	4	55	75.0%	18%	9%	80	94	4.0	25.0%	7.0%	4.8%	$13	10.5%	0%	$140	500
Company B	92%	93%	3	58	72.0%	17%	6%	65	69	6.0	28.0%	11.0%	7.5%	$22	13.0%	5%	$320	1100
Company C	96%	98%	2	62	78.0%	13%	4%	55	74	7.0	22.0%	9.0%	6.1%	$21	11.6%	8%	$285	825
Company D	87%	81%	4	74	78.0%	20%	10%	90	174	2.0	22.0%	2.0%	1.4%	$5	11.4%	–3%	$700	2100
Company E	91%	94%	3	61	71.0%	15%	5%	50	140	4.7	29.0%	14.0%	9.5%	$26	12.0%	7%	$210	760
Median	91.5%	94.5%	3.5	56.5	73.5%	17.5%	6.0%	60	84	5.4	26.5%	10.0%	6.8%	$21	11.8%	5.8%	$303	800
Advantage	95.0%	97.3%	3.0	52.8	71.3%	15.5%	5.3%	51	70	6.8	28.8%	12.4%	8.4%	$25	12.6%	6.9%	$343	1031
Best in class	97.0%	98.0%	2.5	46.0	68.6%	14.0%	4.5%	49	55	7.4	31.4%	13.4%	9.1%	$33	12.9%	7.5%	$525	1600
Gap to median	–6.5%	–3.5%	3.0	–16.5	–7.3%	3.5%	0.0%	–12	–44	–2.5	–7.3%	–2.8%	–1.9%	–$18	–1.0%	–0.8%	–$48	25
Gap to advantage	–3.0%	–0.8%	3.0	–12.8	–5.0%	5.5%	0.7%	–3	–30	–1.1	–5.0%	–0.4%	–0.3%	–$14	–0.2%	0.4%	–$8	256
Gap to best in class	1.0%	0.0%	3.5	–6.0	–2.4%	7.0%	1.5%	–1	–15	–0.4	–2.4%	0.6%	0.4%	–$7	0.1%	1.0%	$175	825

Figure 7.30   MMC year 2 end of year benchmark assessment.

a real disadvantage but the numbers are the numbers. The LEOMS implementation leadership team acknowledged their failure in not dealing holistically with the order fulfillment process and planned to resolve this in the next improvement cycle to be kicked off at the beginning of the new year. MMC's SG&A cost, still the highest among competitors, was not considered a problem as their continued investment in R&D explained most of their gap versus competitors, and cost reduction benefit in SG&A from LEOMS had funded an increase in R&D budgets.

Supply Chain Competitive Strategy Assessment

When their results were reviewed (see Figure 7.31), the entire team was very pleased with the first year of integrating the LEOMS into their operational planning, day-to-day operations, and culture as the only strategic performance

Business strategy aligned supply chain performance		Strategic value discipline – proposition		
		Competitor C	MMC	Competitor B
		Operational excellence	Product leadership	Customer intimacy
		Lowest delivered cost	Technological innovation	Total solutions provider
Customer facing	Delivery reliability	Superior	Advantage	Median
	Responsiveness	Superior	Disadvantage	Advantage
	Adaptability/flexibility	Advantage	Superior	Disadvantage
Internal facing	Cost	Disadvantage	Superior	Median
	Asset management	Advantage	Superior	Median

Figure 7.31   MMC generation 2 supply chain competitive assessment.

attribute that remained a major disadvantage was responsiveness as measured by order fulfillment lead time, and the plan to resolve this in the next year had already been agreed to by the team.

Enabling the Supply Chain

A critical factor in institutionalizing desired operational processes is making sure information technology systems can be configured to operate processes that apply best practices for your company's specific markets' characteristics. Information technology systems must truly enable defined process activities and task in the optimum sequence and timing to really enable effective and efficient execution each day.

This is much easier with today's configurable information technology systems, but that does not automatically mean every system will enable processes perfectly as defined by process work done with SCOR supply chain process improvement methodology. It is critical that organizations define their nonnegotiable process activities and sequences to prospective IT systems suppliers before embarking on system selection so that the right provider partner and system are

---

LEARNED FROM EXPERIENCE

At 3M Brazil, in 1997, we realized our supply chains were not capable of meeting key market segments' needs, and, in fact, we had used generic metrics, which had no meaning to customers for measuring our internal performance. We started a very ambitious program by assigning a full-time team to redesign our business processes. The project turned out to be much bigger than we anticipated as we discovered how many of our commercial business practices affected supply chains. For example, pricing and merchandising practices were impacting supply chain performance and needed to be included in the re-engineering. As a result, nearly two years were spent re-engineering processes before we could begin ERP system implementation in 1999. ERP system modifications were made to institutionalize our new business practices. For example, when it came to inventory allocation, we wanted to make sure every part of our operational model "treated our best customers better," so we modified the allocation process to ensure that orders from our "A" classified customers received the first allocation of inventory." In addition, we placed codes on shipping boxes identifying these customers. Transportation companies were trained so when capacity limitations required a decision about priority, they had the information to make the right choice for 3M preferred customers. From 1995–2000 we significantly outgrew the markets we served and our operating margins increased 30%.

Paul Husby

chosen, and, if needed, require providers to define the additional cost demanded for their system to operate as required to support defined supply chain processes. Money spent up front will be well spent, as otherwise companies frequently have only one of two choices: pay for modifications later when all negotiating leverage is gone, or live with the cost and/or customer dissatisfaction resulting from processes that are not capable of operating as desired because their IT system does not provide the needed enablement.

References

1. SCOR 9.0, Supply Chain Council, 2009.
2. Michael Treacy and Fred Wiersema, *Discipline of Market Leaders*, Boston, MA: Addison Wesley, 1995.
3. Dan Jones and Jim Womack, *Seeing the Whole*, Cambridge, MA: Lean Enterprise Institute, 2003.

Chapter 8

LEOMS Application to Transactional Processes

What Will I Learn?

How to apply LEOMS to transactional processes.

Taiichi Ohno's LEOMS Vision

Although Lean practices have most commonly been applied to supply chain operations, they are equally applicable to all processes and functions in a business. In fact, Taiichi Ohno repeatedly described the Toyota Production System as a management system, which would work in every type of business. He clearly saw the LEOMS goal as elimination of waste everywhere it existed in a business. Over the last three decades, greater understanding of Lean, along with Taiichi Ohno and Shigeo Shingo's genius, has been slowly discovered and understood. While this chapter focuses on supply chain transactional processes, this approach can be applied on transactional processes of all organization functions. As understanding of the LEOMS has grown, practices and principles are being applied to additional functions and industries, expanding the LEOMS's continuous improvement process application resulting in greater improvement impact.

From Manufacturing to Services Industries

A growing trend is applying the LEOMS in service industries. Many industries have published examples of adopting Lean; software development and healthcare are two examples. Many hospitals and hospital groups across the United States are applying Lean to reduce waiting times and increase utilization of doctor's time, equipment, and facilities. Applying Lean is improving quality, reducing cost, and increasing customer satisfaction. Julia Hanna wrote about applying Lean at Wipro, the Indian software development company in her article Bringing "Lean" Principles to Service Industries.[1] Two Harvard researchers found that Wipro's initial application of Lean to their development projects increased efficiency greater than 10% in 80% of the projects. This initial effort grew to 603 Lean projects within 2 years, producing improved productivity and empowered work teams. Their research illustrated five examples of Lean practice application:

1. Using kaizen has altered software development approaches from sequential methods with work moving from one developer to another to iterative approaches of teams completing logical software functionality collaboratively.
2. Sharing of mistakes across development teams to learn and apply these experiences to future projects.
3. Wipro used tools like the system complexity estimator, which compared actual architectures to ideal state architectures, helping teams understand where additional resources would be needed during a project.
4. Value stream mapping was applied to projects, identifying waste of time and effort of projects, leading to increased speed and productivity.
5. Engagement and empowerment energized the organization as all team members, regardless of organizational level, were seeing the bigger picture. This resulted in thousands of software engineers contributing to innovation through problem solving, creating an energized work environment, while increasing productivity and quality. In a 2005 white paper called Lean Software Development, Kumar Desai[2] codified Lean software concepts and practices, making them more easily applied to other organizations. Included in his white paper are identification of software development's seven wastes, nine Lean principles, and how to create a culture of continuous improvement.

Another industry applying Lean to its business is healthcare. David Wessner,[3] president and CEO of Park Nicollet Health Services, St Louis Park, Minnesota, chronicled their Lean journey in a 2005 Minneapolis Start and Tribune newspaper article. Park Nicollet doctors, nurses, and technicians applied standard Lean analysis tools and practices to analyze and radically improve their operations. These included the use of stopwatches, spaghetti diagrams, standard work, 5S, level loading, and rapid changeovers, to name a few. The results are transforming Park Nicollet. The first clinic studied was endoscopy, resulting in doubling the number of patients processed each day. This success was replicated in the cancer center, heart center, urgent care, and wound clinic. The use of standard work applied to surgery resulted in 40,000 fewer instruments being used each month; this simplification also meant fewer errors as the right instrument was always available during surgery. Park Nicollet reported completing 85 rapid process improvement workshops resulting in $7.5 million in savings.

LEOMS Aligns Organizations to Increase Productivity and Quality

All of us have had experiences of working in organizations operating with some level of chaos and cordial anarchy as they tend to exist together (see Figure 8.1). Cordial anarchy is easy to identify simply by getting to know leaders and their organizations as its symptoms quickly show up. Examples include leaders "saying the right things in front of their boss about working with peers," but they only focus on their own stuff and make decisions based on what is good for their organization. Organized chaos is also common as processes and IT systems age without much attention to adapting to real-world process changes occurring because of changes in business environments, customer expectations, and their business processes. Another major symptom of cordial anarchy is superficial collaboration among organizational members. Applying the LEOMS to enterprise business processes requires collaboration as end-to-end processes involving multiple functions builds a recognition that collaboration benefits all functions involved in a process and creates improvement through common project objectives. Cordial anarchy is a

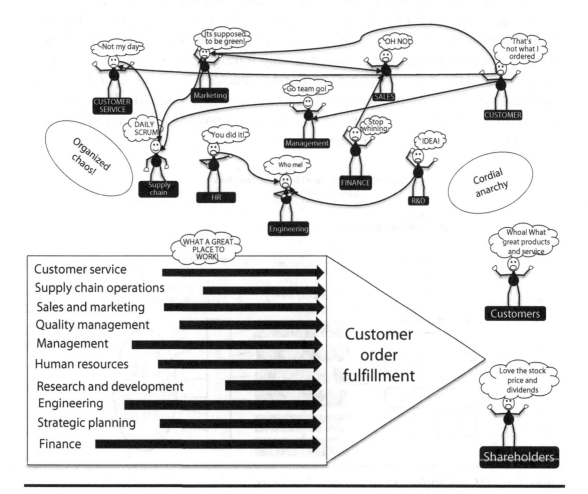

Figure 8.1 Synchronizing organizations.

cultural artifact and a powerful reason why culture change was addressed in Chapter 5 to create collaboration as an attribute in organizations implementing the LEOMS. So, what is a process? A process is a collection of activities that takes one or more inputs and creates an output that has value to the enterprise and its customers.

From the Shop Floor to the Enterprise

The LEOMS is being applied to processes outside of the supply chain, taking the LEOMS from the shop floor to the enterprise (see Figure 8.2). Some U.S. businesses have been on their Lean journey for three decades, and only now is the LEOMS's thinking being broadly understood and applied. LEOMS application in many respects has been restrained by its great success in manufacturing, creating a perception that the LEOMS is to be applied only to shop floor operations. As previously discussed, this was not the founders' intention, and only now are we seeing broader adoption of Lean as an enterprise management system. The common evolutionary application thread of the LEOMS to organizations is that broader and deeper application occurs as greater understanding is gained of the LEOMS and the genius of its founders. Practicing leads to seeing and understanding more about the LEOMS's organic nature as a system. The LEOMS's implementation is a never-ending journey, and it is only through practice and study that deeper understanding of improvement opportunities are discovered and solutions applied. The LEOMS's life cycle is still in its embryonic phase, which speaks powerfully to its future contributions to radically improving competitiveness of businesses, government services, and even NGOs.

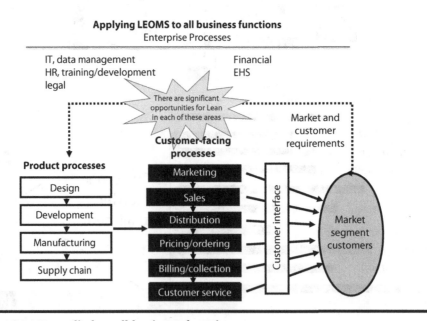

Figure 8.2 LEOMS applied to all business functions.

From Supply Chain Operational Processes to Transactional Processes

Every function, from product development to marketing, customer service, sales, finance, product development, process and facilities engineering, and IT, has processes for getting work done. This means business process mapping and PDCA problem solving can be used to continuously improve their processes.

Increase Marketing Professionals' Value-Added Time

- Are marketing campaigns always effective?
- Are communications with the sales team late, too frequent, or too infrequent?
- Is the sales team provided with timely new product knowledge/expertise to effectively sell the product?
- Are there delays in communicating price increases to the sales force and customers?
- Do pricing errors and price discrepancies exist?

Increase Sales Representatives' Value-Added Time

How much of most sales organizations' effort is value added? One only has to spend windshield time with field salespeople to understand how much non-value-added time exists because of poor communications, lack of credible information, not meeting promises, product quality, order delivery, and paperwork failures. Most sales jobs also have enormous amounts of necessary non-value-added time, such as setting up appointments, preparing for appointments, following up with customer service on existing order status, maintaining multiple customer relationships, doing favors for customers, traveling to and from customer locations, and reporting to headquarters. Realization of these non-value-added activities puts into perspective the real value of the small percentage of work time that sales professionals have available for selling face to face with customers. Applying Lean principles and tools to selling processes will have similar effects as applying them to supply chains. Simply thorough identification of value-added, non-value-added, and non-value-added but necessary work activities alone will provide a treasure chest of opportunities to improve productivity and sales professional job satisfaction. What is the reality regarding the following questions?

- What percent of sales representatives' work time is value-added time?
- Does the organization really support sales representatives, or do sales representatives personally follow critical orders through the entire system because of a lack of confidence?

- How much time are sales representatives spending on administrative tasks and not in the field selling?
- Are sales representatives overly burdened with detailed reporting and reports?
- Are sales representatives not able to follow up on leads in a timely manner?

Increase Customer Service Representatives' Value-Added Time

Most customer services organizations have a mission of supporting customers by providing information and assistance with a goal of increasing customer satisfaction and loyalty. The reality is that most customer service organizations expend most of their effort dealing with failures like getting "real" promise dates from their factory, unmet promise dates, wrong products shipped, or product quality issues. As Lean supply chain operational maturity is achieved, failure causes will be eliminated, allowing customer service operations to become proactive contributors. Their time can be invested in increasing sales, developing prospective customers, and building stronger relationships with existing customers. The application of Lean to customer service processes will eliminate waste and improve speed, thus increasing value-added contribution. The LEOMS applied to customer service processes will lead to resolving important questions about the quality of customer service processes such as

- What percent of customer service representative's time is value added?
- Are customers confused on whom to contact for help?
- Do customers have long wait times to get help?
- Do customers get timely responses to complaints?
- Do customers get the information or assistance they are desiring?

Increase Product Development and Commercialization Processes' Value-Added Time

Lean thinking and tools should also be applied to new product development and commercialization processes. Increasing speed and effectiveness of commercialization will not only have significant cost benefits but also create strategic advantage if new product commercialization cycle times become significantly faster than competitors. In addition, focus on customer value leads development organizations to team up with factory operations to ensure that sufficient process capability exists to reliably and repeatedly produce new designed products. This benefit is strategically valuable to both product innovation leaders and fast followers to sustain their competitive positions. The use of value stream maps to understand end-to-end commercialization processes along with their cycle times, review and approval delays, waiting time, capacity barriers, and skill barriers all provide myriad opportunities to improve speed, effectiveness, and cost of commercialization. LEOMS thinking will challenge the status quo:

- Are product development cycle times too long?
- What causes delays?
- Are launch deadlines missed?
- Why are project milestone dates missed?
- Is the commercialization process inefficient?
- Do newly launched products have quality levels that satisfy customers?

Increase Human Resources' Value-Added Time

- Do frontline team members see human resources as a trusted partner that represents and fights for their interest? How do you know?
- Do you have long cycle times to hire an employee?
- Do employees receive timely feedback and direction?
- Is human resources investing in, coaching, and training frontline team members?

Increase Finance and Accounting Services' Value-Added Time

Finance and accounting operational services are infrastructure processes that benefit from applying Lean; for example, continuously improve processes by applying swim lane process mapping to their processes, having standardized data models and output deliverables, providing automation of self-service report generation, and data analysis. More routine activities such as month end closing along with monthly and quarterly financial reporting can greatly be streamlined to reduce cost and cycle times. Closing financials a couple of days after the end of a quarter means that investors get timely information and minimal resources are being consumed to get it done. Cycle times reduction of financial accounting processes enables operations to respond sooner to any deviations or early negative trends with cause analysis and corrective action. Perfection is having only needed information at the point it is needed and at the moment it is needed, at the lowest possible cost by pursuing the answer to questions such as:

- Are inefficient forecasting processes resulting in multiple forecast revisions?
- Is excessive time spent creating custom reports (which the requesters wait too long for)?
- How do you eliminate disputed customer invoices that delay payment?
- Are internal organizations getting information that they need to manage efficiently and effectively?

Increase IT Professionals and Systems' Value-Added Time

The LEOMS will assist IT organizations to continuously improve by pursing perfection in their processes by answering questions such as the following:

■ Do managers have access to the data they need to run their business—or do they need to fund a project to get the data?

■ Do existing systems provide enablement of best practices and task sequences along with the data needed to support business processes effectively and efficiently?

■ Can customers get information they need directly from your web applications—or do they need to contact a sales/service office to get it?

■ Is there one version of the truth for critical enterprise data (such as financials, product sales, customer accounts, etc.)?

■ Is there a defined system of record for each data element or do users have to hunt for it?

■ Is it easy to determine in advance what systems/processes will be impacted by a planned change to a database?

■ Are different tools or standards used by different teams within the enterprise to perform similar data integration work?

■ Do system maintenance costs increase each time an integration point is added?

■ Are integration activities the constraint (as per TOC) in most projects?

LEOMS thinking and eight wastes are applicable to continuously improve effectiveness and efficiency of systems operations and provide users with improved quality, timeliness, and easier access to data by users. Systems have continued to evolve and typically have more options and flexibility because of modern systems' ease of configuration, but it is still necessary to be sure that system selection and implementation will enable business process practices and tasks defined by the future state process map effectively and efficiently.

- Segment all processes into strategic, core, and infrastructure processes to provide direction and context to operating, improving, and sustaining each process.
- Identify the process owner to participate and own ongoing continuous process improvement
- Map the process:
    - Walk it from the end back through every step, finishing at its first step
    - Define inputs and outputs
    - Interview team members responsible for each process step
- Create the current state process
- Create a future state process
- Create a current state process with improvements to identify process improvement projects required to achieve the future state process
- Determine improvement project priorities by considering
    - Impact on performance
    - Process dependencies—often best to start at the beginning
- Implement processes improvement
- Establish a control plan to maintain the gains

Figure 8.3   Transactional business process improvement.

What Is the Process and Where to Start LEOMS's Business Process Improvement Implementation?

Business processes are sequenced sets of activities executed by business functions to get work done. To be effective and efficient, companies need to have a defined transactional business process improvement methodology that is shared and applied across all business functions. Transactional business process improvement steps are shown in Figure 8.3.

Segment Enterprise Business Processes

All enterprise business processes are not equal in terms of their enterprise value creation and sustainment and for its customers. Processes can be segmented into strategic, core, and infrastructure (see Figure 8.4). MMC, for example, built its business on new and innovative products based on innovative materials and process technologies. This requires a significant investment in scientists, engineers, laboratories, pilot plants, intellectual property lawyers, along with product development organizations focused on market segments and their applications. It also requires a well-trained sales force who can sell their value-added product benefits and technical service organizations to support industrial, commercial, and government customers who use MMC products that require knowledge regarding their applications and use. It is a "high cost" business model, so they must invest significantly to generate new innovative solutions for customers to maintain their loyalty, sustain MMC's growth, and support their brand reputation. Sustaining this model requires continued investment in new technology and acquisition of startup technology companies with proprietary technology that fits solutions for MMC's key market customers. These processes are "strategic processes," and improving these is a high priority to ensure that they are maintaining relevance to their customers by identifying and participating in evolving and future technologies relevant to enterprise markets. This requires investing in resources and infrastructure to develop and enable commercialization of new technologies. A company such

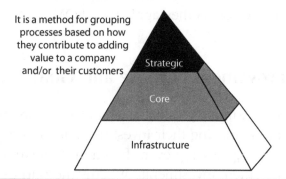

Figure 8.4 What is process segmentation?

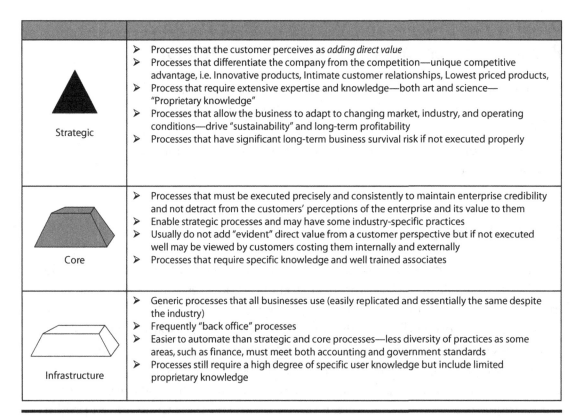

Strategic	➢ Processes that the customer perceives as *adding direct value* ➢ Processes that differentiate the company from the competition—unique competitive advantage, i.e. Innovative products, Intimate customer relationships, Lowest priced products, ➢ Process that require extensive expertise and knowledge—both art and science—"Proprietary knowledge" ➢ Processes that allow the business to adapt to changing market, industry, and operating conditions—drive "sustainability" and long-term profitability ➢ Processes that have significant long-term business survival risk if not executed properly
Core	➢ Processes that must be executed precisely and consistently to maintain enterprise credibility and not detract from the customers' perceptions of the enterprise and its value to them ➢ Enable strategic processes and may have some industry-specific practices ➢ Usually do not add "evident" direct value from a customer perspective but if not executed well may be viewed by customers costing them internally and externally ➢ Processes that require specific knowledge and well trained associates
Infrastructure	➢ Generic processes that all businesses use (easily replicated and essentially the same despite the industry) ➢ Frequently "back office" processes ➢ Easier to automate than strategic and core processes—less diversity of practices as some areas, such as finance, must meet both accounting and government standards ➢ Processes still require a high degree of specific user knowledge but include limited proprietary knowledge

Figure 8.5   How are business processes segmented?

as MMC would have a few selective supply chain processes segmented as strategic with the majority in core and infrastructure segments compared to a company competing as a low-cost supplier who would define many of their supply chain processes as strategic. Figure 8.5 defines attributes of strategic, core, and infrastructure processes. Identifying processes in each of the three segments depends on each enterprise's basis of competition and differentiating competencies. Developing the proposed segmentation for each function is best done by management and a specialist in each function with assistance from their organization's strategic planning leadership. Once a draft is completed, it should be reviewed by functional executives with a final review and approval by the company executive committee. This is a significant investment of time, but it is very important in the long term as segmentation provides enterprise organizations with direction regarding what they should focus on as they apply the LEOMS to their processes.

Continuous Improvement Purpose and Goal

Process segmentation is extremely valuable as it clarifies organization's purpose and priorities related to managing their investments in continuously improving and redesigning processes, allowing every function to make the best choices in their process improvement investments. What is the purpose and focus of improvement process investment in each process segment?

Strategic Processes

Strategic process continuous improvement focuses on enhancing customer perception and experience while executing them at the lowest cost and highest quality. There may be cases when it is important to change the process in a way that increases internal cost but contributes significantly to customers' experiences and/or perceptions of value provided.

Core Processes

Core process continuous improvement focuses on striving for perfection in performance, quality, reliability, waste elimination, increasing adoptability and flexibility to meet changing customer requirements.

Infrastructure Processes

Infrastructure process continuous improvement aggressively focuses on perfect quality and driving down cost. In all cases, the LEOMS's eight wastes are used for opportunity identification to improve quality and reduce cost. The methodology for seeing and documenting the process is swim lane process mapping, the transactional process equivalent of value stream mapping of the supply chain operational processes.

MMC Transactional Process Segmentation

MMC transactional process segmentation (see Figure 8.6) chose to focus on their order fulfillment process as their performance after a year of applying the

Gain executive consensus on strategic processes

		Supply chain	Customer service	Finance
Strategic		• Manage process engineering R&D • New facility location and scale of operations • Procurement strategy • Manage proprietary manufacturing processes	• Manage customer relationship process	• Define and execute the financial strategy and structure • Tax strategy and plan
Core		• Facility design and construction process • Materials requirement planning • Purchase order process • Supplier management process • Physical order fulfillment process • Material receiving and storage process • Cycle inventory process • Transportation supplier selection process • Transportation delivery process • Manage returns and exchanges	• Manage order fulfillment process • Manage back orders • Process customer inquiries • Provide field support and service • Manage recalls • Manage warranties—field support • Evaluate project performance • Calculate order price	• Lead annual budget preparation • Set annual proposed budget goals • Assist functional units in completing their budgets • Provide monthly budget status reports • Manage credit exposure • Invoicing (billing) • Accounts payable • Accounts receivable
Infrastructure		• Manage product return process • Manage inventory data • Manage customer order and demand data	• Maintain customer order data • Manage returns and exchanges • Maintain accurate and timely demand information	• General ledger transactions posting, management, and period closing • Fixed asset accounting • Project accounting • Tax management • Overhead accounting, allocation, and product costing

Figure 8.6 MMC process segmentation.

LEOMS to their supply chain was still putting them at a disadvantage versus their competitors regarding their total order fulfillment lead time. Its negative impact was extremely serious, so it must be resolved as it has affected operational cost, customer service, and inventory levels. Working with customers to improve their entire order fulfillment process allows MMC to add value to their businesses while reinforcing MMC's value as a supplier. Meetings with customers need to be well planned with hard data and relevance to their business. By sharing openly with their customers and exposing their own issues, it opened a constructive dialogue related to how both parties could contribute to improving a process for their mutual benefits.

Preparing Organizations for Continuous Process Improvement

Kicking off transactional process, LEOMS continuous improvement is focused on gaining support and commitment from all organizational levels. This should start with top leadership being engaged to understanding the process, their critical roles, and their commitment to dedicate their best people to lead the training and implementation. Figure 8.7 is an outline of tasks required to prepare an organization to start improving their processes. With continuous improvement process leaders assigned, it is time to prepare organizational departments to begin their continuous improvement journey.

Engaging Department Team Members to Improve Their Processes

Prior to engaging teams in process mapping, it is helpful to engage them in an exercise to help them become enthusiastic about improving their processes. These workshops should include all department team members, as they must own the continuous improvement responsibility for their own processes. A good exercise with department team members is to start a dialogue by identifying current wastes (see Figure 8.8). Allowing a group or logical subgroups to spend time brainstorming about waste in their processes creates team member energy and momentum to move forward with improving their processes. The brainstorming

- Work with top department leadership to assign a process improvement leader
- Explain the purpose of training to all department members
- Explain the various roles needed for sustainable success—total participation
- People doing processes are responsible and empowered to improve it
- Teach them the continuous improvement process
- Use table top simulations to reinforce understanding of continuous improvement

Figure 8.7 Preparing teams to improve their processes.

Transactional business process improvement opportunities

Transactional processes	Opportunities	Process improvement
• Customer service	• Hand-offs	• Assess
• Finance	• Number of touches	• Swim lane process map
• Human resources	• Reworking data	• Current state
• New product introduction	• Searching in multiple	• Future state
• New system introduction	data bases	• Current state improvements
• Order fulfillment	• Duplication of work	• Metric performance
• Product transfer	• Rechecking someone	• Identify waste
• Packaging design	else's work	• Identify solutions
• Sales	• Wait time	• Implement solutions
• Technical service		• Confirm solutions' effectiveness

Figure 8.8 Where will we find the waste?

output also provides facilitators with relevant material for use in training groups on problem solving. Once participants see that there are opportunities and know they are empowered to implement change, they will want to get started with making it happen. The team should be ready to engage in learning how to create process maps (see Figure 8.9). While process maps will eventually be documented on computer applications using Excel, Power Point, or Visio; it is best to start the old-fashioned way with brown paper on a wall, large Post-It notes, and colored markers. The facilitator and his/her team walk upstream through the entire process and document it along with all waste observations. With the process and waste observations documented, it is time to gather the team assigned to kick off the project by sequencing process steps using large Post-It notes placed on brown paper. This will help the group as they start to see where the mapping process is going, making it easier for them to engage.

Transactional business process mapping

- Involve the process owners in the entire process so they are prepared to lead their team to sustain continuous improvement
- Gather and review all related documentation, i.e. business rules, requirements, and documentation.
- Walk the process from the last process step back through every step and finish with first step.
- Define inputs and outputs
- Interview the team member responsible for each process step
    - Introduce yourself and engage team member(s)
    - Explain your purpose for the discussion and their role in process improvement
    - Provide context—benchmark information, VSM's, geo maps
    - Be respectful, thorough, inquisitive, and challenging
    - Use open ended questions
- Document data, notes, and responses to be used as inputs to the current state process
- Document wastes discovered from team member discussions and observations
- Build the process map by engaging people doing the work

Tools	Resources
• Brown paper	• Process team members
• Blue masking tape	• Person to document discussion
• Post-it notes	• Related value stream maps and geo maps
• 4"×6" for process steps	• Geo maps with supplier and key customer information
• 3"×5" for naming waste	• Benchmark information
• Markers	

Figure 8.9 MMC process mapping.

Mapping a process for the first time with a group is messy, so it is better to do it on brown paper first. This enables everyone to be actively involved, writing down wastes and opportunities they saw, and to present them to the group as they place their Post-It notes on the value stream illustrated on the brown paper. It is best to continue with brown paper sessions until processes are well defined and the group is fully engaged in improving their process, so that they can easily relate to it as they move from participants to owners.

Transactional Business Process Mapping

Swim lane process mapping is the best tool for documenting transaction processes involving more than one department and/or customers and suppliers (see Figure 8.10). The swim lane process map captures every process step, sequences and places them in lanes based on the process owner as the overall process moves from left to right in the lanes.

Mapping the process:

1. Functions: list names of each function in a sequential order following the process along with the total value-added time of their process steps.

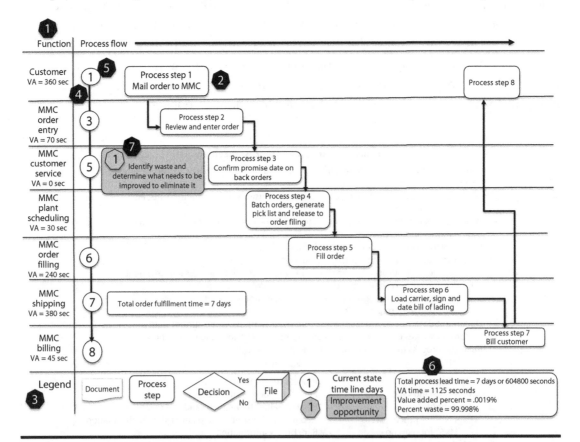

Figure 8.10   Swim lane process map.

2. Process steps should be completed by walking the process upstream from its final step, in this case the customer mailing in their order, through each process step back to the first step documenting process inputs and outputs.

3. Legend contains shapes for typical process activities from information sources such as documents and files to process steps, activities, and decisions. In addition, key information is captured such as total process lead time, improvement opportunities, and a summary value-added percent calculation.

4. Value-added time is the measured amount of time in each process that adds value for the customer.

5. Total process lead time is the clock time from process initiation until it is complete.

6. Process value-added time percentage is calculated by dividing the measured value-added time by total production lead time.

7. Improvement opportunities are identified after a future state process is defined and projects identified that will change the process from its current state to its defined future state.

MMC Order Fulfillment Process Improvement

MMC focused on supply chain operations in their first year of the LEOMS deployment and made good progress, but while they improved their shop floor pick list release to shipment by 50% from 4 to 2 days (see Figure 8.11), unfortunately it was not even noticeable to customers as they still saw total order fulfillment times from customers initiating an order until receipt of between 9 and 12 days (see Figure 8.12). The light came on as it was obvious that they needed to radically shorten process lead time starting from customers submitting orders. Step 1 was to map their order fulfillment process using swim lane process mapping (see Figure 8.13), including their key accounts in the process. MMC had become complacent and had not faced reality; they were losing market share because customers no longer believed that their innovative product prices could be justified because overall service and quality was not competitive with their other suppliers. MMC had chosen to reduce operating expenses and capital investments to maintain margins. The consequences were out-of-date IT systems and poor process equipment operating conditions. Finally, the light came on when they saw benchmarking results showing their weak competitive position confirmed by their best customers. The leadership team took a couple of days to put the shock behind them and was ready to move to right the ship and get back to being the market leadership company they were in the past.

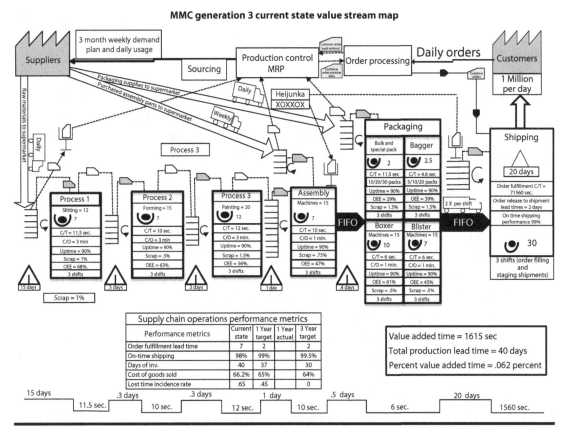

Figure 8.11 MMC VSM current state generation 3.

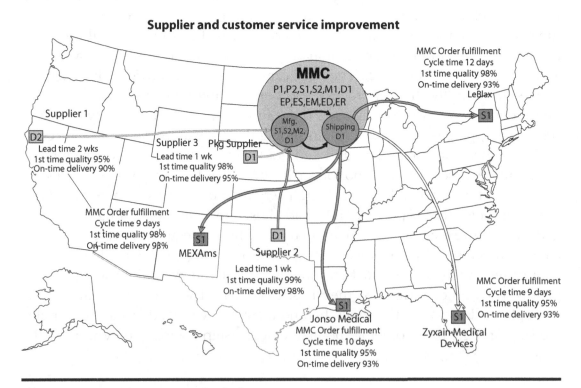

Figure 8.12 MMC year 1 supply chain map.

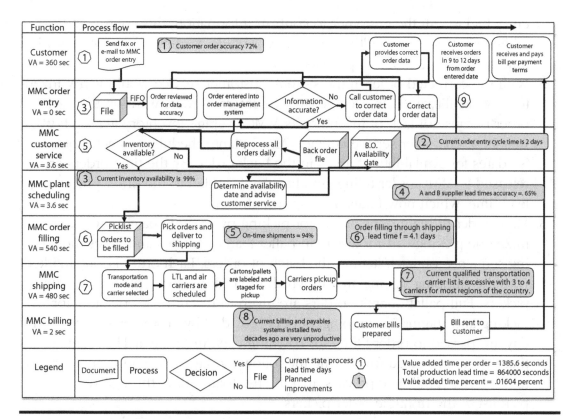

Figure 8.13   MMC order fulfillment process current state.

Improving the Order Fulfillment Process

MMC Current State Order Fulfillment Process

MMC had become complacent believing that their superior product and process technology were all they needed to satisfy customers; they were decades behind because of their lack of investment in their processes and IT systems information. Recognizing this failure, they were motivated to start the implementation of the LEOMS in their order processing and customer service areas, while factory operations were completing their first cycle through the LEOMS seven-wave implementation process. There are seven functions listed on the left side of Figure 8.13 that are involved in the order fulfillment process starting with the customer:

1. MMC's lack of up-to-date information technology meant that their customers sent in their orders using a fax or by e-mail using a scanned PDF document or occasionally by mail; see the top swim lane labeled "Customer" in Figure 8.13. This meant that total time from a customer's order arrival at MMC until it was in their order entry system averaged 1 day, as shown in lead-time days (white hexagon 1).

   With the order in their system, it could be reviewed and edited for completeness and accuracy, but currently accuracy was 72% as shown in the text box connected to the gray-colored hexagon 1. Most of the time, MMC order entry had to call the customer to confirm information; this contributed to adding another day, resulting in a 2-day average current state process lead time through the first two process steps.

2. With the corrected order in the system, customer service would verify inventory availability and, on average, 10% of entered orders went on the back-order because of lack of inventory as MMC average inventory availability was 90%. Customer service contacts MMC plant scheduling to get promise dates for availability of items in the back-order. By the time orders are released to MMC order filling, another day is added to the order processing lead time, which now totals 3 days through step 3.

3. MMC order filling groups orders for picking by creating batches to maximize one picking cart for less than pallet load quantities and notes the picking sequence on each item in the batch of orders. Each order is also marked with the shipping staging area in which the order is to be placed by order pickers. Full pallet quantity orders are given to reach truck operators for picking and staging in shipping. Orders provided by customer service each day are passed to order filling for batching each morning at 11:00 a.m. so that they are released for picking the following day, which adds another day to the order fulfillment process, making the total 4 days through release of pick list to the frontline order pickers.

4. A copy of orders staged at shipping each day is provided to the shipping office for freight carrier selection and creation of a bill of lading, and is then returned to the staging area for team members to restage orders organized by the freight carrier who would deliver them; this antiquated process usually consumed a full day of lead time.

5. The following day, shipments were prepared and restaged for loading.

MMC realized that their very antiquated order processing, order filling, and shipping processes had a completely unacceptable order entry through release to order filling with a total process lead time average of 5 days, and it needed to be reduced to 1 day. With that target in mind, they began by creating an MMC order fulfillment process future state with a target order fulfillment lead time of 2 days.

Creating the MMC Order Fulfillment Process Future State

MMC envisioned a future state with a 2-day order fulfillment lead time (see Figure 8.14) as their future state goal. Achieving this goal would require a radical redesign of their processes and implementation of a new enterprise IT system to enable the new processes. MMC could not wait until a new system was chosen, configured, and implemented; it would take up to 2 years to see the results, so they decided to redesign the processes and hire temporary team members to execute tasks and activities required to achieve the target results of a 2-day order fulfillment cycle time. They were confident this would work.

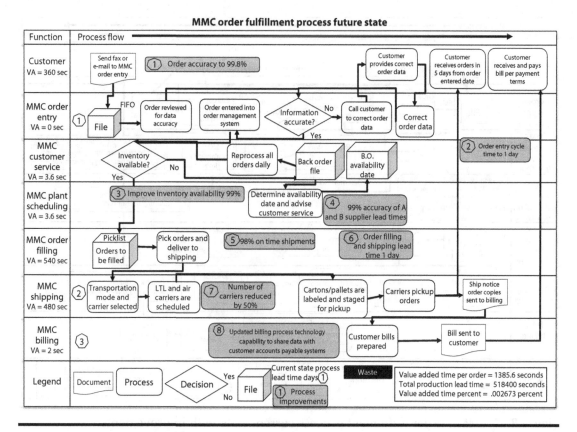

Figure 8.14 Order fulfillment process future state.

Define Required Improvements

MMC's order fulfillment improvement team was selected and challenged with the task of identifying process improvements necessary to achieve future state targets. While MMC's path would require both time and money, it also had a side benefit of proving the proposed solutions prior to investing in IT systems. It would also give them clear process enablement specifications for their selected IT system supplier. The improvement team identified eight significant improvement targets (see Figure 8.15):

1. Improve customer order accuracy from 72% to 99%.
2. Reduce order entry to pick list generation cycle time from the current 5 days to 1 day.
3. Improve supply chain response to achieve a 37-day total production lead time and increase fill rate from 94% to 98%.
4. Establish a process for updating A and B supplier lead times quarterly and ensure that they are correct to reduce the chance of late deliveries because MMC assumed an out-of-date lead time.
5. Improve on-time shipments from 94% to 98%.
6. Reduce order release to shipment lead time from 2 days to 1 day.

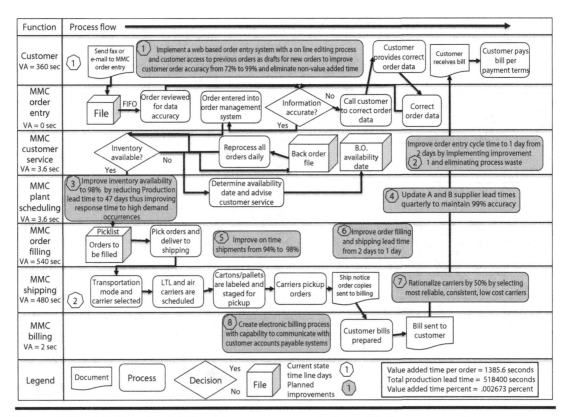

Figure 8.15 MMC order fulfillment process current state with improvements.

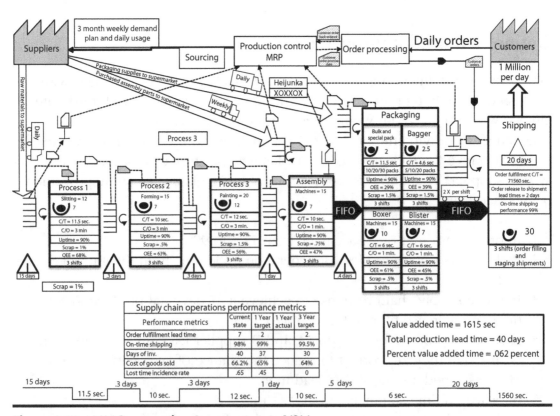

Figure 8.16 MMC generation 3 current state VSM.

7. Rationalize freight carriers by 50% by selecting the most reliable, consistent delivery and lowest cost carriers.
8. Implement electronic billing process with capability for two-way data transfer and communications with customer accounts payable system.

The order fulfillment improvement team had started their work when factory operations were well into their recently completed second improvement cycle, shown in Figure 8.16. The good news for the order fulfillment improvement team from MMC operations was that they had made improvements during the intervening period and already met or are close to meeting some of their targets:

Target 3: Operations had improved line item fill rate to 98% and set 99% as the target for MMC VSM generation 4.
Target 5: On-time shipments achieved 98% and set a 99% target.
Target 6: Reduce order release to shipment lead time from 2 days to 1 day is the operations target for their MMC VSM generation 4.

The order fulfillment team continued to work on their improvement projects as did factory operations, which continued to work on their third cycle through LEOMS waves, as both had challenging goals to reach by year end.

Year-End Review

The year-end review was highly anticipated by both the factory operations team and the order fulfillment team as they were fully engaged and motivated by what they had learned and applied in the past year.

Factory Operations Report

The director of factory operations introduced his team to company executives attending the annual review, who would be approving the targets (for the coming year) and 3-year strategic goals. The vice president of supply chain operations then used five slides in his presentation, starting with MMC generation 3 current state VSM (see Figure 8.16) laying out the targets for the current year. He then moved quickly to the MMC generation 4 VSM current state (see Figure 8.17) to focus on the improvements made and strategic and operational benefits they had achieved. The presentation was then focused on generation 4 VSM's supply chain performance metrics showing the great progress MMC had made in their 3-year LEOMS implementation journey (see Figure 8.18). The next slide was MMC's benchmark performance generation 4 (see Figure 8.19), which had just been completed and updated showing the enormous improvement in their market position versus peers and competitors. The last slide presented by the vice president of operations was supply chain competitive strategy assessment

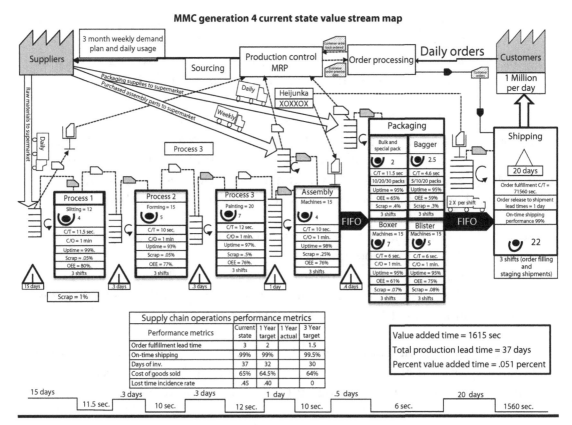

Figure 8.17 MMC generation 4 VSM current state.

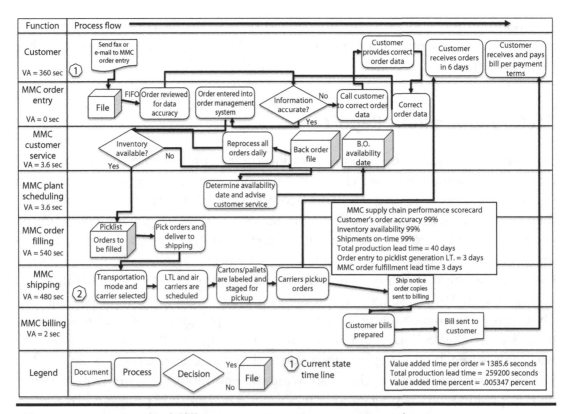

Figure 8.18 MMC order fulfillment process current state generation 2.

MMC benchmark performance generation	On-time shipment performance percent	Line fill rate percent	Order fulfillment leadtime	Total production lead time	Cost of goods sold percent	SGA cost percent	Warranty/returns cost percent	Cash to cash cycle in days	Days of inventory	Asset turns	Gross margin	Operating income percent	Net income percent	Value added per employee ($ 000)	Return on capital employed	3 yr CAGR Percent	Revenue ($ 000,000)	Number of employees
MMC	99%	99%	2	37	64.5%	21%	4%	44	37	7.2	35.5%	14.5%	9.9%	$48	12.0%	7%	$380	775
Company A	83%	95%	4	55	75.0%	18%	9%	80	94	4.0	25.0%	7.0%	4.8%	$13	7.5%	0%	$140	500
Company B	92%	93%	3	58	72.0%	17%	6%	65	69	6.0	28.0%	11.0%	7.5%	$22	13.0%	5%	$320	1100
Company C	96%	98%	2	52	78.0%	13%	4%	55	74	7.0	22.0%	9.0%	6.1%	$21	11.6%	8%	$285	825
Company D	87%	81%	4	74	78.0%	20%	10%	90	174	2.0	22.0%	2.0%	1.4%	$5	2.4%	-3%	$700	2100
Company E	91%	94%	3	61	71.0%	15%	5%	50	140	4.7	29.0%	14.0%	9.5%	$26	12.0%	7%	$210	760
Median	91.5%	94.5%	3.0	56.5	73.5%	17.5%	5.5%	60	84	5.4	26.5%	10.0%	6.8%	$21	11.8%	6.0%	$303	800
Advantage	95.0%	97.3%	2.3	52.8	71.3%	15.5%	4.3%	51	70	6.8	28.8%	13.3%	9.0%	$25	12.0%	7.0%	$365	1031
Best in class	97.5%	98.0%	2.0	44.5	67.8%	14.0%	4.0%	47	53	7.1	32.3%	14.3%	9.7%	$37	12.5%	7.5%	$540	1600
Gap to median	-7.5%	-4.5%	-1.0	-19.5	-9.0%	3.5%	-1.5%	-16	-47	-1.9	-9.0%	-4.5%	-3.1%	-$27	-0.2%	-1.0%	-$78	25
Gap to advantage	-4.0%	-1.8%	-0.3	-15.8	-6.7%	5.5%	-0.3%	-7	-33	-0.5	-6.7%	-1.3%	-0.8%	-$23	0.0%	0.0%	-$15	256
Gap to best in class	-1.5%	-1.0%	0.0	-7.5	-3.3%	7.0%	0.0%	-3	-16	-0.1	-3.3%	-0.2%	-0.2%	-$11	0.5%	0.5%	$160	825

Figure 8.19   MMC benchmark performance generation 4.

Business strategy aligned supply chain performance		Strategic value discipline—proposition		
		Competitor C	MMC	Competitor B
		Operational excellence	Product leadership	Customer intimacy
		Lowest delivered cost	Technological innovation	Total solutions provider
Customer facing	Delivery reliability	Superior	Superior	Median
	Responsiveness	Superior	Superior	Advantage
	Adaptability/flexibility	Superior	Superior	Median
Internal facing	Cost	Disadvantage	Superior	Median
	Asset management	Median	Advantage	Median

Figure 8.20   Supply chain competitive strategy assessment.

(see Figure 8.20). The group in the room could not resist celebrating and congratulating each other. When the celebration ended, the CEO went to the front to wrap up the meeting.

Review Wrap-Up

The company CEO reminded the top leadership group of how far they had come once they took an honest look in the mirror and accepted that their culture created by the incredible company founders based on technological innovation leadership, had become complacent with a false sense of superiority. The initial benchmarking and customer visits were the source of their awakening and soul searching that started them on the LEOMS journey, a journey that had resulted in a new belief that their legacy would be their contribution to extending the founders' vision and values through adoption of the LEOMS, which had revitalized their value proposition that had built a great company. Adoption of the LEOMS, if sustained, will prevent future leadership generations from falling into the same trap of arrogance and complacency that had threatened to permanently damage MMC. After thanking all the participants, he excused them to have a meeting with his executive committee. When the room was cleared of all except his executive committee, he asked the secretary to call the meeting to order as they had something very important that needed to be discussed and to decide on the right action to take. The CEO explained how proud he was of their progress and that one of its consequences was that the operations now had more team members than they needed. Operations had redeployed people to handle material delivery routes and support continuous operations through shift changes, lunch, coffee breaks, and meetings, but still had up to 50 team members without a defined work assignment. The room became quiet as they awaited the CEO's proposal to resolve this issue. The CEO stated how proud he was of all team members who were committed to their LEOMS journey, and they deserved to be treated both respectfully and generously. He then asked the HR vice president to explain the early retirement buyout plan he approved. The HR vice president described an extremely generous buyout plan that would be offered to all factory production workers over 58 years old as they needed about 40% of them to accept the buyout to get their headcount aligned with current operations reality. The executive committee voted unanimously to approve the buyout as described as they felt confident this would send a message to all company team members that they are valued and respected regardless of their position in the company.

References

1. Julia Hanna, *Bringing "Lean" Principles to Service Industries*, Harvard Business School, 2007.
2. Kumar Desai, *Lean Software Development, Perfect Project White Paper Collection*, 2005.
3. David Wessner, Park Nicollet hospital's Lean initiative, *Minneapolis Start and Tribune*, 2005.

Appendix I: Book Resources for Practitioners

Suggested Library for the LEOMS implementation professionals and/or teams:

A. Great resources for shop floor supervisors, group leaders, team members, and maintenance:

1. *5S for Operators*, Productivity Press Development Team, Productivity Press, 1996.
2. *Kanban for the Shop Floor*, Productivity Press Development Team, Productivity Press, 2002.
3. *Just-in-Time for Operators*, Productivity Press Development Team, Productivity Press, 1998.
4. *Mistake-Proofing for Operators*, Productivity Press Development Team, Productivity Press, 1997.
5. *Quick Changeover for Operators*, Productivity Press Development Team, Productivity Press, 1996.
6. *TPM for Every Operator*, Productivity Press Development Team, Productivity Press, 1992.

B. Training books/manuals for Lean implementation:

7. Rick Harris, Chris Harris, and Earl Wilson, *Making Materials Flow*, The Lean Enterprise Institute, 2003.
8. Mike Rother and Rick Harris, *Creating Continuous Flow*, The Lean Enterprise Institute, 2001.
9. Art Smalley, *Creating Level Pull*, The Lean Enterprise Institute, 2004.
10. Mike Rother and John Shook, *Learning to See*, The Lean Enterprise Institute, 2003.
11. Rick Harris, *Developing a Lean Workforce*, Productivity Press, 2007.
12. Tom Jackson, *Hoshin-Kanri for the Lean Enterprise*, Productivity Press, 2006.
13. Durward K. Sobek and Art Smalley, *Understanding A3 Thinking*, 2008.
14. Johh Shook, *Managing to Learn: Using the A3 Process to Solve Problems, Gain Agreement, Mentor and Lead*, The Lean Enterprise Institute, 2008.

15. Mark Hammel, *Kaizen Event Field Book*, Society of Manufacturing Engineers, 2010.

16. Donald A. Dinero, *Training within Industry*, Productivity Press, 2005.

17. Arthur Hill, *The Encyclopedia of Operations Management*, Clamshell Beach Press, 2010.

C. Taiichi Ohno and other Japanese authors:

18. Taiichi Ohno, *Workplace Management*, McGraw-Hill, 2013.

19. Taiichi Ohno, *Total Production System—Beyond Large-Scale Production*, Productivity Press, 1988.

20. Satoshi Hino, *Inside the Mind of Toyota*, Productivity Press, 2006.

21. Masao Nemoto, *Total Quality Management*, Prentice Hall, 1987.

22. Kaoru Ishikawa, *Guide to Quality Control*, Asian Productivity Organization, 1982.

23. Masaaki Imai, *Gemba Kaizen*, McGraw-Hill, 2012.

D. Books related to Toyota and the Toyota Production System:

24. Jeffery Liker, *The Toyota Way*, McGraw-Hill, 2004.

25. Jeffery Liker and David P. Meier, *Toyota Talent*, McGraw-Hill, 2007.

26. Jeffery Liker and Michael Hoseus, *Toyota Culture*, McGraw-Hill, 2008.

27. Mike Rother, *Toyota Kata*, McGraw-Hill, 2010.

28. Jeffery Liker and James Morgan, *The Toyota Production Development System*, Productivity Press, 2006.

E. Culture and organizational change management:

29. Kim S. Cameron and Robert E. Quinn, *Diagnosing and Changing Organizational Culture*, John Wiley & Sons, 2011.

30. John P. Kotter, *Leading Change*, Harvard Business Review Press, 2012.

31. Dan S. Cohen, *Heart of Change Field Guide*, Harvard Business Review Press, 2005.

32. Jon Miller, Mike Wroblewski, and Jaime Villafuerte, *Creating a Kaizen Culture*, McGraw-Hill, 2014.

F. Business leadership and competitive advantage (learnings from looking back at best-selling business book and article premises, conclusions, and reality):

33. Robert Waterman, *The Renewal Factor*, Bantam, 1987.

34. Charles Collin and Jerry Porras, *Built to Last*, Harper Business, 1994.

35. Michael Tracey and Fred Wiersema, *Discipline of Market Leaders*, Addison Wesley, 1995.

36. Jennifer Reingold and Ryan Underwood, *Was 'Built to Last' Built to Last?*, Fast Company, 2004.

37. Steven J. Spear, *High-Velocity Edge*, McGraw-Hill, 2009.

38. Thomas J. Peters and Robert H. Waterman, *In Search of Excellence*, HarperCollins, 1982.
39. Gary Collins, *Why Do Large Companies Fail*, Analytics, 2012.

G. Lean enterprise product development:

40. James M. Morgan and Jeffrey K. Liker, *The Toyota Product Development System: Integrating People, Process and Technology*, Productivity Press, 2006.

Appendix II: Taiichi Ohno Core Lean

Toyota Production System **(Productivity Press, 1988)—Taiichi Ohno Quotes**

Chapter	Page	Topic	Subtopic	Quotation
2	17	Kaizen	PDCA	The Toyota Production System has been built on the practice and evolution of this scientific approach. By asking why five times and answering it each time, we can get to the real cause of the problem, which is often hidden behind more obvious symptoms. In a production operation, data is highly regarded, but I consider facts to be even more important.
1	9	A total management system		A total management system is needed that develops human ability to its fullest capacity to best enhance creativity and fruitfulness, to utilize facilities and machines well, and to eliminate all waste.
1	6	Autonomation		Toyota emphasizes autonomation, machines that can prevent problems "autonomously." A machine automated with a human touch is one that is attached to an automatic stopping device as well as various safety devices, fixed position stopping, the full-work system and poke-yoke-fool proofing systems to prevent defective products. In this way human intelligence, or a human touch is given to a machine.

(Continued)

Chapter	Page	Topic	Subtopic	Quotation
2	41	Autonomation—defect detection		To ensure 100 percent defect free products, we must set up a system that automatically informs us if any process generates defective products.
1	4	Basis of the Toyota Production System		The basis of the Toyota Production System is the absolute elimination of waste. The Two Pillars needed to support the system are: Just-in-time and Autonomation (or automation with a human touch).
3	57	Complacency cuts hope of any progress		The way we currently operate, the production line has a fairly high operational rate and fairly low defect rate, therefore, as a whole, things seem to be proceeding reasonably. If we allow ourselves to feel this way, we cut off any hope for progress and improvement.
3	69	Concept	Ninjutsu	I frequently say management should be done not by arithmetic but by Ninjutsu, the art of invisibility. As children, we watched Ninjutsu tricks at the movies, like the hero suddenly disappearing. As a management technique, however, it is something very rational. To me it means acquiring management skills, by training. No goal, no matter how small can be achieved without adequate training.
2	25	Flow		The work arena is like a track relay race, there is always an area where the baton is passed. If the baton is passed well the total time can be better than the individual times of the four runners.

(*Continued*)

Chapter	Page	Topic	Subtopic	Quotation
2	19	Identify waste	Preliminary step	The preliminary step toward application of the Toyota Production System is to identify wastes completely: • Waste of overproduction • Waste of time on hand (waiting) • Waste of transportation • Waste of processing • Waste of stock on hand (inventory) • Waste of movement • Waste of making defective products • Waste of meaningless jobs and team member talent
3	48	Information technology		At Toyota, we do not reject the computer, because it is essential in planning production leveling procedures and calculating the number of parts needs daily. We use the computer freely, as a tool, and try not to be pushed around by it. We want information only when we need it.
3	59	Inventory	The worst waste	Inventory has to be moved and neatly stacked. If these movements are regarded as "work," soon we will be unable to tell waste from work. In the Toyota Production System, this phenomenon is called the waste of overproduction, "Our Worst Enemy," because it helps hide other wastes.
2	19	Kaizen	True efficiency	True efficiency improvement comes when we produce zero waste and bring the percentage to 100 percent. Since, in the Toyota Production System, we must make only the amount needed, and manpower must be reduced to trim excess capacity and match the needed quantity.

(*Continued*)

Chapter	Page	Topic	Subtopic	Quotation
2	29	Kanban	Automatic nerve of the production line	Kanban, in essence, becomes the automatic nerve of the production line. Based on this, production workers start work by themselves and make decisions about overtime by themselves. The Kanban system also makes clear what must be done by managers and supervisors.
2	30	Kanban	The function and rules of Kanban	Functions of Kanban and Rules of Use: 1) Provides pick-up or transport information—Rule–Later process picks up the number of items indicated by the Kanban at the earlier process. 2) Provides production information—Rule–Earlier process produces items in the quantity and sequence indicated by the Kanban. 3) Prevents overproduction and excessive transportation—Rule–No items are made or transported without a Kanban. 4) Serves as a work order—Rule–Always attach a kanban Card to the goods. 5) Prevents defective products by identifying the process making the defectives—Rule–Defective products are not sent on to the subsequent process. The result is 100% defect free goods. 6) Reveals existing problems and maintains inventory control—Rule–Reducing the number of kanban increases their sensitivity.
2	42	Kanban	System continuous improvement	It should be the duty of those working with kanban to keep improving it with creativity and resourcefulness without allowing it to become fixed at any stage.

(Continued)

Chapter	Page	Topic	Subtopic	Quotation
2	43	Kanban	Chain conveyor application	When a variety of parts are carried on a chain conveyor, indicators designating the parts needed are attached to the conveyor hangers at regular intervals to eliminate any mistake in the type of part, quantity, or time it is required. Thus, by installing a means of conveying only the parts indicated, smooth delivery and withdrawal of needed parts can be achieved. Production leveling is maintained by circulating the part-indicators with the conveyor.
3	51	Kanban	Inventory quantity control practice	To cope with a constantly fluctuating market, the production line must be able to respond to schedule changes. In reality, however, the information system and production constraints make change difficult. An important characteristic of kanban is that within certain limits it makes fine adjustments automatically.
2	33	Kanban (system)	Prerequisite	The Toyota Production System is the production method and the kanban system is the way it is managed. Unless one completely grasps this method of doing work so that things will flow, it is impossible to go right into the kanban system when the time comes. For this tool to work fairly well, the production processes must be managed to flow as much as possible. This is really the basic condition. Other important conditions are leveling production as much as possible and always working in accordance with standard work methods.

(*Continued*)

Chapter	Page	Topic	Subtopic	Quotation
2	27	Kanban (system)	Kanban card	Kanban is a card with information about materials pick-up, transfer and production. It is used to request and authorize replenishment of materials within Toyota and with cooperating firms. It is the autonomic nerve of the production system. Based on it, production workers start by themselves, and make their decisions concerning overtime.
3	48	Lean system adaptability		I think a business should have reflexes that can respond instantly and smoothly to small changes in the plan without having to go to the brain. If a small change in a plan must be accompanied by a brain command to make it work, for example, the production control department issuing order changes and plan change orders, the business will be unable to avoid burns or injuries and will lose great opportunities. Building a fine-tuning mechanism into the business so that change will not be felt as change is like implanting a reflex nerve inside the body. This nerve system is built on having necessary information when needed and Toyota scheduling system it's pacemaker scheduling.
1	9	Lean applies to all businesses		This production system will work for any type of business.
1	4	Lean pillars		The basis of the Toyota Production System is the absolute elimination of waste. The two pillars of the system are, Just in time and Autonomation (automation with a human touch).

(*Continued*)

Chapter	Page	Topic	Subtopic	Quotation
0	xi	Lean principle	Autonomation	TPS is much more than Kanban and Judoka. It's making a factory operate just like the human body operates. The autonomic nervous system responds even when we are asleep. It is only when a problem arises that we become conscious of our bodies. Then we respond by making corrections. The same thing happens in a factory. We should have a system in a factory that automatically responds when problems occur.
0	xv	Lean system relevance in the information age		The Toyota Production System however is not just a production system. I am confident it will reveal its strength as a management system adapted to today's era of global markets and high-level computerized information systems.
2	31	Management commitment		I have good reason for emphasizing the role of top management in discussing the rule of Kanban. There are many obstacles to implementing the rule that the later process must take what it requires from the earlier process when it is needed. For this reason, management commitment and strong support are essential to the successful application of the first rule.
2	36	Management commitment	Overcoming resistance	When I was, rather forcefully, urging foremen in the production plant to understand kanban, my boss received a considerable number of complaints. They voiced the feeling that this fellow Ohno was doing something utterly ridiculous and should be stopped. This put the top manager in a difficult position at times, but even then he must have trusted me. I was not told to stop and for this I am grateful. In 1962, kanban was adopted company-wide; it earned its recognition.

(Continued)

Chapter	Page	Topic	Subtopic	Quotation
0	xiii	Motivation for creating Lean		The Toyota Production System evolved out of need. Certain restrictions in the marketplace required the production of small quantities of many varieties under conditions of low demand, a fate the Japanese automobile industry had faced in the postwar period.
1	3	Motivation for creating Lean		I still remember my surprise at hearing that it took nine Japanese to do the job of one American. President Toyoda was saying we should catch up in three years. If we could eliminate the waste, productivity should rise by a factor of ten. This idea marked the start of the present Toyota Production System.
1	14	Motivation for creating Lean		After WWII, our main concern was how to produce high quality and we helped the cooperating firms in this area. After the oil crisis we started teaching outside firms how to produce goods using the Kanban system. Toyota employees would go and help.
1	13	Necessity is the mother of invention		I strongly believe that "necessity is the mother of invention." Even today, improvements at Toyota plants are made based on need. The key to progress in production improvement, I feel, is letting the plant people feel the need.
2	25	Need for practice		The most important point in common between sports and work is the continuing need for practice and training.
2	34	Partnering with suppliers		Toyota employees would go help so people from nearby firms understood the system early although they faced resistance in their companies. Today it is a pleasure to see the fruit of this work.

(Continued)

Chapter	Page	Topic	Subtopic	Quotation
2	18	PDCA practice	5 Whys	In a production plant operation, data are highly regarded, but consider facts to be even more important. When a problem arises, if our search for the cause is not thorough, the actions taken can be out of focus. This is why we repeatedly ask why. This is the scientific basis of the Toyota System.
1	8	Practice	Autonomation	Autonomation, on the other hand, performs a dual role. It eliminates over production, an important waste in manufacturing, and prevents the production of defective products. To accomplish this, standard work procedures, corresponding to each player's ability, must be adhered to at all times. When abnormalities arise—that is, when a player's ability cannot be brought out—special instructions must be given to bring the player back to normal. This is an important duty of the coach.
1	13	Practice	Work cell design	Rearranging the machines on the floor to establish a production flow eliminated the waste of storing parts. If also helped us achieve the "one operator, many machines" system and increased production efficiency two and three times.
1	7	Practice	Autonomation	Autonomation changes the meaning of management as well. An operator is not needed while the machine is working normally. Abnormalities will never disappear if a worker always attends to a machine and stands in for it when an abnormality does occur. Expanding the rule, we established that even in a manually operated production line, the workers themselves should push the stop button to halt production if any abnormality appears.

(*Continued*)

Chapter	Page	Topic	Subtopic	Quotation
2	18	Practice	Value stream mapping	Improving efficiency only makes sense when it is tied to cost reduction. Look at the efficiency of each operator and of each line. Then look at the operators as a group, and then at the efficiency of the entire plant (all the lines). Efficiency must be improved at each step and at the same time, for the plant as a whole.
2	21	Practice	Standard work	Standard work sheets are posted prominently at each workstation. They are a means of visual control, which is how the Toyota Production System is managed. High production efficiency has also been maintained by preventing the recurrence of defective products, operational mistakes, and accidents, and by incorporating workers ideas. All this has been possible because of the inconspicuous standard work sheet.
2	22	Practice	Training new team members	It should take only three days to train new workers in proper work procedures. When instruction in the sequence and key motions is clear, workers quickly learn to avoid redoing a job or producing defective parts.
2	23	Practice	Standard inventory	Standard inventory refers to the minimum intra-process work in process needed for operations to proceed. This includes items mounted on machines.
2	23	Practice	Work flow sequence	The term work sequence means just what it says. It does not refer to the order of processes along which products flow. It refers rather to the sequence of operations, or the order of operations in which a worker processes items: Transporting them, mounting them on machines, removing them from machines, and so on.

(Continued)

Chapter	Page	Topic	Subtopic	Quotation
2	26	Practice	Supermarkets	A supermarket is where a customer can get what is needed, at the time needed. In principle, however, the supermarket is a place where we buy according to need. From this we got the idea of viewing the earlier process as a kind of store. This helped us approach our just-in-time goal and, in 1953, we actually applied the system in our machine shop at the main plant.
2	31	Practice	SMED	SMED—Single minute exchange of die. Trying to make only the items withdrawn also means changing the setup more often unless the production line is dedicated to one item. Consequently, shortening setup times and reducing lot sizes becomes necessary.
2	37	Practice	Heijunka-level scheduling	Because fluctuation in production and orders at Toyota's final automobile assembly have a negative impact on all earlier processes, Toyota must avoid these negative cycles and must lower the peaks and raise the valleys in production as much as possible so that the flow is smooth. This is called leveling, or load smoothing, ideally should result in zero fluctuation at final assembly line, or the last process.
2	39	Practice	Lot size of one is the ultimate goal	The Toyota Production System, however, requires leveled production and the smallest lots possible even though it seems contrary to conventional wisdom. To respond to the dizzying variety in product types, the die must be changed often. Consequently, setup procedures must be done quickly.

(*Continued*)

Chapter	Page	Topic	Subtopic	Quotation
2	40	Practice	Level loading	A definite production plan can be divided by the number of days to level the number of cars to be made per day. On the production line, even finer leveling must be done to let sedans or coupes flow continuously during a fixed time interval.
3	49	Practice	Pace maker scheduling	In the Toyota Production System, the method of setting up this daily schedule is important. During the last half of the previous month, each production line is informed of the daily production quantity for each product type. At Toyota, this is called the daily level (takt time). On the other hand the daily sequence schedule is sent to only one place—the final assembly line. This is a special characteristic of Toyota's information system. In other companies, scheduling information is sent to every production process.
3	50	Practice	Cars being assembled carry needed information	Too much information induces us to produce ahead and can also cause a mix-up in sequence. Items might not be produced when needed, or too many might be made, some with defects. Eventually, it becomes impossible to make a simple change in the production schedule. In business, excess information must be suppressed. Toyota suppresses it by letting the products being produced carry the information.
3	60	Practice	Takt time	Takt is the length of time, in minutes or seconds, it takes to make one piece of the product. It must be calculated in reverse from the number of pieces to be produced. It is calculated by dividing the operable time per day by the required number of pieces per day. Operable time is the length of time that production can be carried out per day.

(*Continued*)

Chapter	Page	Topic	Subtopic	Quotation
3	60	Practice	OEE metric	The operating rate is the production record of a machine based on its fulltime operational capacity.
Glossary of terms	129	Practice	Eight wastes	The preliminary step toward application of the Toyota Production System is to identify wastes completely: • Waste of overproduction • Waste of time on hand (waiting) • Waste of transportation • Waste of processing • Waste of stock on hand (inventory) • Waste of movement • Waste of making defective products • Waste of Ninjutsu—skills and knowledge training—the invisible Toyota magic
3	60	Practice	Operable machine rate	The operable rate refers to the availability of a machine in operable condition when it is need. The ideal rate is 100 percent.
0	xiii	Principle	Respect for humanity	The most important objective has been to increase production efficiency by consistently and thoroughly eliminating waste. This concept and the equally important respect for humanity are foundations of the Toyota Production System.
0	xiv	Principle	Customer value	We are now unable to sell our products unless we think ourselves into the very heart of customers, each of whom has different concepts and tastes. Today, the industrial world has been forced to master in earnest the multi-kind, small quantity production system.

(*Continued*)

Chapter	Page	Topic	Subtopic	Quotation
1	8	Principle	Judoka	Autonomation, on the other hand, performs a dual role. It eliminates overproduction, an important waste in manufacturing, and prevents the production of defective products. To accomplish this, standard work procedures, corresponding to each player's ability, must be adhered to at all times. When abnormalities arise, that is what a player's ability cannot be brought out; special instruction must be given to bring the player back to normal. This in an important duty of the coach.
2	23	Principle	Teamwork	In modern industry, harmony among people in a group, as in teamwork, is in greater demand than the art of the individual craftsman.
2	18	Principle	Total system efficiency	Improving efficiency only makes sense when it is tied to cost reduction. Look at the efficiency of each operator and of each line. Then look at the operators as a group, and then at the efficiency of the entire plant (all the lines). Efficiency must be improved at each step and at the same time, for the plant as a whole.
2	20	Principle	Go see	I have always believed in going to the plant first principle. The time that provides me with the most vital information about management is the time I spend in the plant, not in the vice president's office. I have been unable to separate myself from the reality found in the plant, not in the vice president's office.
2	22	Principle	Standard work	Standard work sheet combines materials, workers and machines to produce efficiently. It clearly lists the three elements of the standard work procedure as: 1. Cycle time 2. Work sequence 3. Standard inventory

(Continued)

Chapter	Page	Topic	Subtopic	Quotation
2	24	Principle	Teamwork	At work things do not necessarily run smoothly just because areas of responsibility have been assigned. Teamwork is essential.
2	40	Principle	Shared general use equipment	In keeping market diversification and production leveling in harmony, it is important to avoid the use of dedicated facilities and equipment that could have more general utility. More effort is needed to find the minimum facilities and equipment required for general use, utilizing all available knowledge to avoid undermining the benefits of production.
2	41	Principle	Process stability	Efforts to thoroughly stabilize and rationalize the processes are the key to successful implementation of automation. Only with this foundation can production leveling be effective.
3	52	Principle	Learn from production interruptions	The role of fine adjustments is not only to indicate whether a schedule change is a "go" or a "temporary stop" but also to enable us to find out why a stop occurred and how to make the fine adjustments necessary to make it go again.
3	53	Principle	Continuous cost reduction	Manpower reduction at Toyota is a company-wide activity whose purpose is cost reduction. Therefore, all considerations and improvement ideas, when boiled down, must be tied to cost reduction. Saying this in reverse, the criterion for all decisions is whether cost reduction can be achieved.
3	57	Principle	Cost prevention	At Toyota, we go one step further and try to extract improvement from excess capacity. This is because, with greater production capacity, we don't need to fear new cost.

(Continued)

Chapter	Page	Topic	Subtopic	Quotation
3	58	Principle	Raise value-added work percent	Manpower reduction means raising the ratio of value-added work. The ideal is to have 100 percent value-added work. This has been my greatest concern while developing the Toyota Production System.
3	61	Principle	Labor linearity	Now, suppose that market demand, that is, the required number of production, drops to 100 or 90 pieces per day. What happens? If we continue to make 120 pieces a day because of our improved efficiency, we will have 20 to 30 pieces left over daily. This will increase our material and labor expenses and result in a serious inventory problem. How can we improve efficiency and still reduce costs? The problem is solved by improving the process so that eight workers can produce 100 parts, and if 90 parts are needed, seven workers should be used. Increasing efficiency by increasing production while the actual demand remains the same is called an "apparent increase of efficiency."
3	62	Principle	Tortoise and the Hare	In a plant where defined numbers actually dictate production, I like to point out that the slower but consistent tortoise cause less waste and is much more desirable that the speedy hare who races ahead and then stops occasionally to doze.
3	64	Principle	Maintaining existing equipment is always best for cost	My conclusion is that if adequate maintenance has been done, replacement with a new machine is never cheaper, even if maintaining that older one entails some expense. If we do decide to replace it. We should realize that we have either been misled by our calculations and made the wrong decision or that our maintenance program has been inadequate.

(Continued)

Chapter	Page	Topic	Subtopic	Quotation
3	65	Principle	We are surrounded by a harsh environment	Business management must be very realistic. A vision of the future is important but it must be down-to-earth. In this age, misreading reality and its ceaseless changes can results in an instant decline in business. We are indeed surrounded by a harsh environment.
3	66	Principle	Partnering	Only by working as partners with the cooperating firms is it possible to perfect this system. The same is true in improving the character of management. Toyota alone cannot achieve the goal if the cooperating firms do not work together.
3	67	Principle	Worker savings	In our company, we use the term "worker saving" instead of "labor saving." How can we increase production with fewer workers? If we consider this question in terms of the number of worker days, it is a mistake. We should consider it in terms of the number of workers. The reason is that the number of workers is not reduced even with a reduction of .9 worker days.
3	68	Principle	Work cell design should facilitate teams and teamwork	In the Toyota Production System, we frequently say, "do not make isolated islands." If workers are sparsely positioned here and there among the machines, it appears as if there are few workers. However, if a worker is alone there can be no teamwork. Even if there is only enough work for one person, five or six workers should be grouped together to work as a team. By providing an environment sensitive to human needs, it becomes possible to realistically implement a system that employs fewer workers.

(Continued)

Chapter	Page	Topic	Subtopic	Quotation
3	70	Principle	Management by Ninjutsu	We call the Toyota Production System, management by Ninjutsu. If you look up the word engineer in an English Dictionary, you will find "technologist," while in Japanese, its meaning uses the character for "art." Analyzing this character, you will find it is created by inserting the character "require" into the character "action." So, art seems to be something that requires action.
1	15	Principle	Use of resources	A revolution in consciousness is indispensable. Industrial society must develop the courage, or rather the common sense, to procure only what is needed when it is needed and in the amount needed. This requires what I call a revolution in consciousness, a change of attitude and viewpoint by business people. We must understand these situations in-depth before we can achieve a revolution in consciousness.
2	37	Principle	System flexibility	While the traditional planned mass-production system does not respond easily to change, the Toyota Production System is very elastic and can take the difficult conditions imposed by diverse market demands and digest them. The Toyota system has the flexibility to do this.
4	79	Principle and practice	Go see	Stand on the production floor all day and watch, you will eventually discover what has to be done; I cannot emphasize this too much.
4	87	Principle and practice	Go see	Find a subject to think about, stare at an object until a hole is almost bored though it, and find out its essential nature. Stand and watch a neighborhood grandmother's hand loom for a shola day. This was how Toyoda Sakichi was inspired and tracked down the facts.

(Continued)

Chapter	Page	Topic	Subtopic	Quotation
2	44	Process stability	Requirement of kanban	Kanban must work effectively to maintain just in time in the plant and for kanban to be effective, stabilization and production leveling are indispensable conditions. Some people think, however, that kanban can be used only to manage parts processed in daily stable quantities, but this is a mistake. Others think kanban cannot be used without a steady withdrawal pattern of parts. This is also wrong thinking.
3	54	Purpose	TPS eliminates waste and enhances productivity	The Toyota Production System is a method to thoroughly eliminate waste and enhance productivity. In production, "waste" refers to all elements of production that only increase cost without adding value; excess people, inventory and equipment.
0	xiii	Respect for humanity	Team members employed to think	The Toyota style is not to create results by working hard. It is a system that says there is no limit to people's creativity. People don't go to Toyota to "work" they go there to "think." Why not make the work easier and more interesting so that people do not have to sweat?
1	14	Respect for humanity		In the Japanese system, operators acquire a broad spectrum of production skills that I call manufacturing skills and participate in building up a total system in the production plant. In this way, the individual can find value in working.
2	20	Respect for humanity	Use all of people's talents	Eliminating wasteful and meaningless jobs enhance the value of work for workers.
3	52	Respect for humanity	Train and empower team members to cope with change	But as long as we cannot accurately predict the future, our action should change to suit changing situations. In industry, it is important to enable production people to cope with change and think flexibly.

(Continued)

Chapter	Page	Topic	Subtopic	Quotation
0	ix	Response to question: What are you doing now?		All we are doing is looking at the timeline, he said, "from the moment the customer gives us an order to the point when we collect the cash." And we are reducing that time line by removing the non-value added wastes.
2	21	Standard work	No standard work—no kaizen	"Where there is no standard there can be no Kaizen." Taiichi Ohno We have eliminated waste by examining available resources, rearranging machines, improving machining processes, installing autonomous systems, improving tools, analyzing, transportation methods, and optimizing the amount of materials at hand for machining. High production efficiency has also been maintained by preventing the recurrence of defective products, operational mistakes, and accidents, and by incorporating workers' ideas. All of this is possible because of the inconspicuous standard work sheet.
2	21	Standard work	Scope and purpose	Standard work sheets effectively combine materials, workers, and machines to produce efficiently. At Toyota this procedure is called standard work combination.
2	35	Success requires patience and persistence		It took 10 years to establish kanban at the Toyota Motor Company. Although it sounds like a long time, I think it was natural because we were breaking in totally new concepts. It was, nonetheless, a valuable experience.

(Continued)

Chapter	Page	Topic	Subtopic	Quotation
3	46	System flexibility	Change adaptability	I think a business should have reflexes that can respond instantly and smoothly to small changes in the plan without having to go to the brain. If a small change in a plan must be accompanied by a brain command to make it work, for example, the production control department issuing order changes and plan change orders, the business will be unable to avoid burns or injuries and will lose great opportunities. Building a fine-tuning mechanism into the business so that change will not be felt as change is like implanting a reflex nerve inside the body. This nerve system is built on having necessary, information when needed and Toyota scheduling system (pacemaker scheduling).
1	7	Teamwork and collaboration		What is the relationship between just in time and automation with a human touch, the two pillars of the Toyota Production System? Using the analogy of baseball teams, automation corresponds to the skill and talent of individual players while just in time is the teamwork involved in reaching an agreed upon objective.
2	22	Teamwork	Builds harmony	Carrying out the standard work methods in the cycle time helps worker harmony grow.
2	24	Teamwork	Key to winning	The key to winning is teamwork.

(*Continued*)

Chapter	Page	Topic	Subtopic	Quotation
4	77	Toyota Production System pillars		The two pillars of the Toyota Production System are autonomation and just-in-time. Autonomation was taken from the ideas and practice of Toyoda Sakichi. The Toyota-type autonomous activated loom, which he invented, included a device to automatically stop the machine should any one of many warp threads break or the weft thread run out. Therefore, it is essential that equipment be stopped immediately if there is a possibility of defects.
1	14	Waste	Overproduction	There is no waste in business more terrible than overproduction. Why does it occur? It also leads to an inventory of defects, which is a serious business loss.

Appendix III: Cameron and Quinn's Culture Change Methodology Value

Our approach to diagnosing and changing organizational culture offers six advantages:

1. It is *practical*. It captures key dimensions of culture that have been found to make a difference in organizations' success.
2. It is *efficient*. The process of diagnosing and creating a strategy for change can be accomplished in a reasonable amount of time.
3. It is *involving*. The steps in the process can include every member of the organization, but they especially involve all who have a responsibility to establish direction, reinforce values and guide fundamental change.
4. It is both *quantitative and qualitative*. The process relies on quantitative measurement of key cultural dimensions as well as qualitative methods including stories, incidents, and symbols that represent the immeasurable ambience of the organization.
5. It is *manageable*. The process of diagnosis and change can be undertaken and implemented by a team within the organization—usually the management team. Outside diagnosticians, culture experts, or change consultants are not required for successful implementation.
6. It is *valid*. The framework on which the process is built not only makes sense to people as they consider their own organization but is also supported by extensive empirical literature and underlying dimensions that have a verified scholarly foundation.

In other words, we do not claim that ours is the single best approach, but we do consider it a critically important strategy in an organization's repertoire for changing culture and improving performance.[1]

Reference

1. Kim S. Cameron and Robert E. Quinn, *Diagnosing and Changing Organizational Culture*, p. 24, Hoboken, NJ: Jossey-Bass, 2011.

Appendix IV: Benchmark Data Sources

Once supply chain metrics are identified, benchmarking can proceed. Financial benchmarking of publicly traded competitors and market peers is most commonly done using their 10K reports available from various benchmark services such as Hoovers and Forbes. Data of privately held competitors are not reported publicly and therefore not included in available data sources. Data are also available from free sources but only at the company level, so comparison to multidivision competitors is not possible from these sources. The company's chief financial officer can provide these benchmark comparisons from available sources. Financial benchmarking builds a business case by gaining insight into supply chain improvement opportunities. Operational benchmarking data gathering may require multiple sources to get a comprehensive picture. The American Productivity and Quality Council (APQC), a free service for Supply Chain Council members, is an excellent source of benchmarking data. APQC provides industry-level data but has limited capability to provide data below the industry level. Other data sources include trade associations, published articles and reports, benchmarking services, and customers. Customers normally evaluate suppliers using common criteria. They compare market peers to standards established by the best supplier performances, so not having direct competitors in a benchmarking process does not diminish its value, as the goal is satisfying customers.

Appendix V: Validation of the LEOMS's Value

Five Demonstrable Reasons Why the LEOMS Is the Best Operational Management System

The LEOMS has no equal in delivering increased value to customers and investors making a compelling argument as to why every company should aggressively embrace the Lean enterprise business model. These five deliverables supported by 30 methods or processes are validated in this book.

1. **Delivers the business's value proposition through**
   1. Creating a logical focus on business processes in each of the three segments (strategic, core, and infrastructure processes)
   2. Improving performance, quality, and speed of strategic processes
   3. Improving cost, quality, performance, and speed of core processes, prioritizing those with the greatest contribution to strategic processes
   4. Improving cost, quality, and speed of infrastructure processes
   5. Explicitly identifying and differentiating organizational competencies, facilitating assessment, improvement, and investment to increase strategic competitive advantage
   6. Defining the value proposition explicitly through identification of differentiated competencies and strategic processes (differentiating competency identification and process segmentation)
   7. Linking business plan-driven targets and goals to a value proposition and value discipline (Hoshin Kanri)
   8. Prioritizing organizational focus on the vital few goals (Hoshin Kanri)
   9. Ensuring complete cross-function goal and priority alignment (top-down and bottom-up; Hoshin Kanri)
   10. Applying a planning process that builds balanced goals: internal–customer, strategic–operational, long-term–short-term (Hoshin Kanri)

2. Builds added value from suppliers to customers by

11. Setting end-to-end supply chain goals (supplier to customer value stream targets)
12. Including all participants in operational planning (Hoshin Kanri)
13. Expecting all supply chain participants to contribute to improvement goals (the LEOMS's problem solving)

3. Implements best practices and encourages benchmarking because the LEOMS

14. Is a complete operational system (Lean Enterprise Operational Management System)
15. Contains leading practices that are applied as a part of the LEOMS's implementation (LEOMS improvement practices)
16. Drives for perfection and looks for better practices (Lean principles)

4. Creates customer value as a result of

17. Building strategic capabilities through prioritizing the focus on strategic processes and differentiating competencies (Lean enterprise process segmentation)
18. Scheduling operational planning and review cycles supporting timely adaptation to market changes (LEOMS reviews)
19. Flexibly designing operations to quickly adapt to market changes (Lean reviews and Hoshin Kanri)
20. The LEOMS's ability to automatically adopt to volume and mix variability with minimum cost penalty (Lean value streams)

5. Drives continuous improvement forever as it

21. Holds all levels of leadership accountable (Lean operational reviews)
22. Establishes annual improvement targets (Hoshin Kanri)
23. Constantly increases customer and shareholder value (continuous improvement process)
24. Provides a common language and tools (Lean operational management system model, tools and processes)
25. Keeps improvement gains (standard work and tiered process audits)
26. Applies scientific problem solving (Lean PDCA)
27. Creates and drives toward an end state vision (Lean value proposition and Hoshin Kanri)
28. Utilizes an end-to-end business process assessment (value stream mapping)
29. Builds a culture where associates do and improve their job (Lean associate role)
30. Engages the entire organization (only operational associates add value)

Index

Page numbers followed by f indicate figures.

Printed in the United States
by Baker & Taylor Publisher Services